← → C ⌂ 🗋 inspirehep.net/search?p=find+a+gong,+jeh+tween

# INSPIRE

HEP :: HE

find a gong, jeh tween

Brief format ▼

find | "Phys.Rev.Lett.,105" ... more

Sort by:

Display results:

| latest first ▼ | desc. ▼ | - or rank by - ▼ | 25 results ▼ | single list ▼ |

**HEP** 1. **Superunified Theory: The Foundations Of Science**
Jeh-Tween Gong. 1984. 100 pp.

## Calculating Alpha: fine structure constant

$$\alpha = \frac{e^2}{2\epsilon_0 hc} = \frac{1}{137.035999679}$$

Beta = 1/alpha = 64 ( 1 + first order mixing + sum of the higher order mixing)

$\qquad$ = 64 (1 + 1/Cos A(2) + .00065737 + ...)

$\qquad$ = 137.0359 ...

A(2) is the Weinberg angle, A(2) = 28.743 degrees
The sum of the higher order mixing = 2(1/48)
[(1/64) + (1/2)(1/64)^2 + ...+(1/n)(1/64)^n +...]

$\qquad$ = .00065737 + ...

See: Nature's Manifesto:
https://tienzengong.files.wordpress.com/2019/02/5th-natures-manifesto.pdf

# SUPER UNIFIED THEORY

## THEORY

-THE FOUNDATIONS OF SCIENCE-

JEH-TWEEN GONG

Library of Congress Catalog Card
number 84-90325

International Standard Book Number
ISBN 0-916713-01-6

Printed in the U.S.A.

# PREFACE

Science is the study of the natural world with the purpose of discovering the underlying laws that describe not only the current state of the universe but its evolution as well. Today we rely on the interplay between experiment and theory to probe and examine nature so as to obtain rational models that will lead us to an understanding of the laws of nature.

However, there is no ordinary laboratory likely to be able to attain the temperatures and densities at which the big bang took place. It is also impossible to probe Planck's units of time and distance experimentally directly. Because experiment can offer no guidance, Super Unified Theory (SUT) is speculative. It combines three component theories: Elementary Particle Physics, Cosmology and the Topology of mathematics. And it has quantized gravity and geomentized quanta at the same time.

The goal of physics is to describe nature. Physics has made great progress in describing nature only because it has put aside all attempts to answer the greater problem of explaining nature. However, the goal of this book is to explain nature. A true unified theory has to be able to explain all natural phenomena. And this is the first book attemping just to do that. I might not able to change all physicists' view point now, but I am certain that my idea will be accepted for the

3

generations to come. Galileo had given up his theory once, but the earth has continued to move.

I strongly believe that a theory that is mathematically elegant deserve to be studied even if it does not seem to be able to be verified by experiments in the near future. And I hope that the publication of this book will offer a new direction for physics.

This book involves many years of study and research, during which time Gwendolyn, my wife, has helped me by her encouragement and invaluable assistance. My greatest appreciation goes to her.

Jeh-Tween Gong

Dayton, Ohio
December, 1983

# CONTENTS

# CONTENTS

# CHAPTER ONE

## Introduction of Super Unified Theory

### --- Comparison of unified theories

Many unified theories have been proposed, such as SU(5), SO(10), etc. However, all of them are gauge field theory. These theories are supported by one philosophical concept, one irrational faith and amazingly sucessful history.

This philosophical concept was inplanted into all physicists' brains by Einstein's work. Every system or concept has to be observable or measurable. Even space and time have to be operationally defined. Therefore, he added an artificial condition ($ds^2$ = invariance) which is merely logical in his general theory. With this mandated additional condition, every field has to be a gauge field. And all Lagrangians have to be renormalizable. The conclusion is that a theory must make specific numerical predictions which can then be compared to experimental results. That after all is what science is all about.

The irrational faith for the symmetry principle is really coming from our instinct. The classical mechanics does possess a good symmetry property. It becomes even more important in the quantum mechanics. There is no one able to make us abandon it --- not P-violation, not CP-violation, not even Tviolation. When we eyewitness a spantenous symmetry breaking, our only desire is to construct a higher symmetry scheme --- supersymmetry or super .... supersymmetry.

For the reason of gauge invariance, all functions have to follow the unitary transformation. For the reason of symmetry invariance, we have to utilize the orthogonal similarity transformation. Therefore, we are searching for "Groups", SU(2), SU(3), SU(5), SU(16), SU(3) x SU(2) xU(1), ..., etc. When one group's symmetry is broken, then we try to find a higher dimensional group. From a mathematical view point, any symmetry can be spantenous broken by any arbitrarily introduced perturbation. Now, we are looking for the evidence of spontenous supersymmetry breaking which was caused by the supergravity. It is amazing that we are always able to prove what we set out to prove. It is exactly as if we could always find a particular coin we were searching for in a coin bag which contains more kinds of coins than we could ever possibly imagine. Therefore, we will have a hard time to determine which theory is valid, SU(5) or SO(10), or ...

There is no one able to deny the achievement of the gauge field theory. However, they still have a lot of difficulties. The potential of supergravity has a hidden sector. Even though the fate of our universe is still debatable, the cause of the birth (before big bang) of this universe is unimaginable for the gauge theory. Some one has assumed that a finite size universe is coming from "NOTHING", then it goes through a quantum tunnel. This is a very good assumption. And it can be explained by the concept of Unilogy (see chapter nine). In Chapter V, the supersymmetry is a limit condition which exists   at the point "inf U". However, any degrees of dimension of symmetrical image can be constructed (see chapter VI). And the symmetry breaking scheme is as follows.

$$G(\text{inf } U) \dashrightarrow \ldots , \ldots \dashrightarrow G(n)$$
$$\dashrightarrow G(n - 1) \dashrightarrow \ldots \dashrightarrow G(1) \dashrightarrow SU(3) \times U(1)em$$

inf U is a "Self" (see chapter VII). The total energy of universe(s) is zero (see chapter V). Only inf U can visualize the explosion of the big bang because inf U is massless. The universe is created by borrowing the energy from the next universe (or previous one) through the quantum tunnel.

Obviously, Einstein's defination for time and space is no longer valid (or useful) at the point of the birth of our universe. With the concept of Topology and Unilogy, the gauge and symmetry invariance are no longer the necessary conditions for how this universe is constructed. The only necessary condition is the topological and unilogical invariance. For the particle decay, it can be easily explained by using the edge equage. I used a slightly different form in Chapter II. However, the nonconservation of vacutron and bracket are the direct consequence of the topological properties. In gauge field theory, the monopole is described as the topological defects which appears during the inflationary phase. This kind of monopole not only has a difficulty on cosmology but also violates the topological invariance. In SUT, monopole is necessary for completing the two poles dynamics. Also, there are two kinds of monopole.

SUT has not only unified all physical forces, but also pointed out that topology and unilogy are the unified law. However, the continued development of gauge theory will still enrich our knowledge.

# CHAPTER TWO

## Color forces, particles decay and QED

For the known properties of matter, 24 fundamental particles (quarks and leptons) are needed to represent them all. These 24 particles can be identified with three independent quantum numbers -- Charges, Quark Colors, and Generations (Genecolors). In this paper, I briefly discussed the principles and definations of color forces. Then, I demonstrated how color forces govern some decay processes. And, I derived the QED assumption by using the concept of color forces.

I. Principles and definations.
 A) Defination.
 1) F-particle (fundamental particle) --- leptons and quarks. How can I name Tau lepton and b-quark as fundamental particles when they are always decaying into other leptons and quarks? I have introduced a concept of "bracket". Every bracket possesses three colored seats. The first seat is red. The second seat is yellow. The third seat is blue. Every F-particle has to possess one and only one bracket.
 2) Particle ---- Baryon and meson. Composited particle at least possess two brackets. Meson has two and baryon has three.
 3) Prequark ---- the subparticle of the F-particle which does not possess any bracket. It can only sit on the seat of the bracket. There are two kind of prequarks.

A (Angultron) --- carry +1/3 unit of charge.
-A (anti-Angultron) --- carry -1/3 unit of charge.
V (anti-Vacutron) --- zero charge.
-V (Vacutron) --- zero charge.


B) Colors -- there are two type of fundamental colors and two type of pseudo-colors.

1) F-color (quark color) -- it is determined by the colored seat (red, yellow or blue) of the bracket.

2) Genecolor (generations) -- use the symbol 1, 2, 3 to identify these colors. This color is always carried by the third seat of the bracket.

3) Tempocolor (Temporary color) -- prequark does not possess colors by itself. However, prequark does have power to change F-particle's color. I call this power "Tempocolor".

4) Pseudo anti-F colors.

$u$ = (A, A, V1) has blue F-color.
$\overline{d}$ = (V, V, A1) has blue F-color.

However, these two blue F-colors are pseudo anti-color to each other. They can actually form a pseudo-F-colorless particle such as $\pi^+$ = (u, $\overline{d}$).


C) The unique way to represent every quark and lepton with 3 and only 3 variables -- charge, F-color and genecolor.

FOR QUARK,

|  | Red | Yellow | Blue |
|---|---|---|---|
| u = | (V, A, A1) | (A, V, A1) | (A, A, V1) |
| c = | (V, A, A2) | (A, V, A2) | (A, A, V2) |
| t = | (V, A, A3) | (A, V, A3) | (A, A, V3) |

|  | Anti-R | Anti-Y | Anti-B |
|---|---|---|---|
| d = | -(A, V, V1) | -(V, A, V1) | -(V, V, A1) |
| s = | -(A, V, V2) | -(V, A, V2) | -(V, V, A2) |
| b = | -(A, V, V3) | -(V, A, V3) | -(V, V, A3) |

FOR LEPTON (colorless)

$e^- = -$ (A, A, A1)     $\nu_e = -$ (V, V, V1)

$\mu^- = -$ (A, A, A2)     $\nu_\mu = -$ (V, V, V2)

$\tau^- = -$ (A, A, A3)     $\nu_\tau = -$ (V, V, V3)

Positive bracket carries ordinary colors and negative bracket carries anti-colors. The F-color is identif-ied by the seats' color of the single (alone) prequark. Example, for (V, A, A1), V is the single prequark and it sit on the first seat which is red, therefore, (V, A, A1) has red F-color. Its genecolor is clearly defined by the subscripts 1 on third seat.

D) Color forces principles.

1) Color force priority I ---- The larger priority number the weaker the force is.

Genecolor (T1 force) -- the strongest color force.

F-color (T2 force) -- a stable color force.

Tempocolor (T3 force) -- a perturbating color force.

Pseudo anti-color (T4 force) -- an assistant force.

2) Color force priority II -- the color neutralization priority in the same type of color.

C1 force -- color, anti-color (original) neutralization

C2 force ---- complementory ( red + yellow + blue ) neutralization.

C3 force ----- color, anti-color (complementary ) neutralization.

3) Complementary rule are valid for all colors.

4) Conservation laws.

a) Angultron has to be conserved. But Vacutron can be created from vacuum and disappeared into the vacuum.

b) F- color and genecolor have to conserve accord-ing to complementary rule.

c) Bracket is not conserved.

5) Symbols.

R --- Red.          Y --- Yellow.          B --- Blue
Ra -- anti-R          Ya -- anti- Y.          Ba -- anti-B.
PaR - Pseudo Ra,  Pay - Pseudo Ya,  PaB -Pse. Ba.
Now, I can use the above principles to describe the
quark, lepton, and meson decay.

II. Tau decay.

$$\tau^+ \dashrightarrow \mu^+ + \nu_\mu + \nu_\tau$$

$$\dashrightarrow e^+ + \nu_e + \nu_\tau$$

Tau has genecolor 3. According to color complementary rule, and the conservation law of Angultron, then,

$$A3 \left\{ \begin{array}{l} \underline{\quad\quad} A1 \\ \cdot V1 \\ V2 \left\{ \begin{array}{l} V1 \\ V3 \end{array} \right. \end{array} \right. \quad \text{or} \quad A3 \left\{ \begin{array}{l} \underline{\quad\quad} A2 \\ V1 \left\{ \begin{array}{l} V2 \\ V3 \end{array} \right. \end{array} \right.$$

Therefore,

$$\tau^+ \left\{ \begin{array}{l} A \longrightarrow A \\ A \longrightarrow A \\ A3 \longrightarrow A2 \end{array} \right\} \mu^+$$

(F-colorless)

$$\left. \begin{array}{l} V \longrightarrow V \\ V \longrightarrow V \\ \longrightarrow V2 \end{array} \right\} \nu_\mu$$

$$\left. \begin{array}{l} V \longrightarrow V \\ V \longrightarrow V \\ \longrightarrow V3 \end{array} \right\} \nu_\tau$$

Very clearly, Angultron is conserved. F-color
and genecolor are also conserved according to color
complementary rule. However, Vacutron is not conserved. Muon and Tau decay are pure genecolor
processes.

III. $\pi^o$ decay and its branching ratio.

$$\pi^0 \dashrightarrow r \ r \ \text{(photons)} \sim 90\%$$

$$\dashrightarrow e^+ \ e^- \ r \qquad \sim 1\%$$

$\pi^0$ decay is a pure F-color force decaying process.

A) C1 model for $\pi^0$ decay.

$$\pi^0 \left( \bar{d} \begin{Bmatrix} A & V & V \\ V & A & V \\ Vl & Vl & Al \end{Bmatrix} + \begin{Bmatrix} -A & -V & -V \\ -V & -A & -V \\ -Vl & -Vl & -Al \end{Bmatrix} d \right)$$

$$\Rightarrow \left[ \begin{Bmatrix} A & -A \\ V & -V \\ Vl & -Vl \end{Bmatrix} + \begin{Bmatrix} V & -V \\ A & -A \\ Vl & -Vl \end{Bmatrix} + \begin{Bmatrix} V & -V \\ V & -V \\ Al & -Al \end{Bmatrix} \right]$$

$$\Rightarrow 3 \left[ 2 \ (V, \ V, \ V) \right] \Rightarrow 2 \ (V, \ V, \ V)^* = r \ r$$

The colored quark annihilation will produce two photons for the reason of momentum conservation law. The notation of photon is (V, V, V) which is F-colorless and genecolorless. The annihilation process will cancel out the Angultron, F-color and genecolor. (V, V, V)* is a composited photon.

B) C2 model for $\pi^0$ decay.

$$\pi^0 \begin{cases} \bar{d} \begin{Bmatrix} A & V & V \\ V & A & V \\ Vl & Vl & Al \end{Bmatrix} \longrightarrow \begin{Bmatrix} A \\ A \\ Al \end{Bmatrix} + \\[4mm] d \begin{Bmatrix} -A & -V & -V \\ -V & -A & -V \\ -Vl & -Vl & -Al \end{Bmatrix} \rightarrow \begin{Bmatrix} -A \\ -A \\ -Al \end{Bmatrix} + \end{cases} 2 \begin{Bmatrix} V & -V \\ V & -V \\ Vl & -Vl \end{Bmatrix}$$

$$\Rightarrow (A, \ A, \ Al) + (-A, \ -A, \ -Al) + (V, \ V, \ V)^*$$

$$= e^+ \ e^- \ r$$

C) I have not discussed the C3 force. I will use iso-

lated neutron as an example to discuss the relationship of all three of them. The following is the isolated
neutron color forces structure.

$$\xrightarrow{\quad \text{C2 force} \quad}$$

$$
\begin{array}{ccc}
R & Y & B
\end{array}
$$

$$
u \left\{
\begin{array}{ccc}
V & A & A \\
A & V & A \\
Al & Al & Vl
\end{array}
\right\} \quad (R\text{-}Y\text{-}B)
$$

$$
\begin{array}{ccc}
Ya & Ba & Ra
\end{array}
$$

$$
d \left\{
\begin{array}{ccc}
-V & -V & -A \\
-A & -V & -V \\
-Vl & -Al & -Vl
\end{array}
\right\}
$$

$$
\begin{array}{ccc}
Ba & Ra & Ya
\end{array}
\quad (Ra\text{-}Ya\text{-}Ba)
$$

$$
d \left\{
\begin{array}{ccc}
-V & -A & -V \\
-V & -V & -A \\
-Al & -Vl & -Vl
\end{array}
\right\} \quad \text{complementary}
$$

(C3 force — left margin, pointing downward)

The Cl force is almost 100 times stronger than C2
force. The C2 force is stronger than C3 force. The
C2 and C3 are constructive forces. They are the
forces to bind F-particles together. On the other
hand, Cl is a destructive force.

IV. $\pi^+$ decay.

$$\pi^+ \dashrightarrow \mu^+ \nu_e \quad 100\%$$

$$\dashrightarrow e^+ \nu_e \quad \text{less than tenth}\%$$

$\pi^+$ can be represented as follow,

$$
\begin{array}{ccc}
R & Y & B
\end{array}
$$

$$
u \left\{
\begin{array}{ccc}
V & A & A \\
A & V & A \\
Al & Al & Vl
\end{array}
\right\} \quad \text{F-color}
$$

$$
\bar{d} \left\{
\begin{array}{ccc}
A & V & V \\
V & A & V \\
Vl & Vl & Al
\end{array}
\right\} \quad \text{PaF-color}
$$

$\pi^+$ { F-colorless

It seems to be that $\pi^+ \dashrightarrow e^+ \nu_e$ should be the dominant process.

$$
\pi^+ \begin{cases}
u(B) \begin{cases} A & \longrightarrow & A \\ A & \longrightarrow & A \\ V1 & \longrightarrow & A1 \end{cases} e^+ \\[2em]
\bar{d}(PaB) \begin{cases} V & \longrightarrow & V \\ V & \longrightarrow & V \\ A1 & \longrightarrow & V1 \end{cases} \nu_e
\end{cases}
$$

However, the T4 force (pseudo anti-color force) is just too small compared with the genecolor force.

$\pi^+$ decay is a mixed color force (F-color and gene-color) decaying process. According to genecolor complementary rule.

$$
\left.\begin{matrix} V1 \\ V1 \\ (\ ) \end{matrix}\right\} \rightarrow V2 \text{ or } V3 \qquad , \qquad \left.\begin{matrix} A1 \\ V1 \\ (\ ) \end{matrix}\right\} \rightarrow A2 \text{ or } A3
$$

Very obviously, $\pi^+$ can not decay into a Tau because of the energy conservation law. Therefore, $\pi^+$ can only decay into two brackets which are formed by V2 and A2. And this genecolor process can be described as two steps.

STEP ONE - neutrino forming process (F-color force).

$$
u \begin{cases} A \dashrightarrow (\ ) & V \dashleftarrow V \\ A \dashrightarrow (\ ) & V \dashleftarrow V \\ V1 \dashrightarrow V1 & (\ ) \dashleftarrow A1 \end{cases} d
$$

STEP TWO -- genecolor process.

$$
\left.\begin{matrix} \dashleftarrow (\ ) & V \\ \dashleftarrow (\ ) & V \\ V1 & V1 \end{matrix}\right\} \dashrightarrow \left\{\begin{matrix} V \\ V \\ V2 \end{matrix}\right\} \nu_\mu
$$

Genecolor force is so strong, there is almost always no $\nu_e$ left over. The Moun creation is as follow (one example).

$$
\left\{\begin{matrix} A & V \\ A & V \\ V1 & A1 \end{matrix}\right\} \longrightarrow \left\{\begin{matrix} A \\ A \\ A2 \end{matrix}\right\} \mu^+
$$

The above discussion are only some special case of this process. T1 + T2 force is so strong that $\pi^+ \dashrightarrow$ $\mu^+ \nu_\mu$ is the dominant process.

V. b-quark decay.

b-quark decay is also a mixed color force process. In the $\pi^+$ decay process, different colored quarks interact together to produce some end products which are F-colorless. Also, the single colored b-quark can not decay just by itself. B-quark decay is also a group action. I can write b-quark as follow.

$$b \left\{ \begin{array}{ccc} Ra & Ya & Ba \\ -A & -V & -V \\ -V & -A & -V \\ -V3 & -V3 & -A3 \end{array} \right\} \text{F-colorless}$$

The b-quark decay is not exactly the same as the Tau decay. The following process has to be permitted.

$$-A3 \to \left\{ \begin{array}{c} +V2 \\ -A1 \end{array} \right\} \quad \text{or} \quad \left\{ \begin{array}{c} -V \\ +A2 \end{array} \right\}$$

The same for -V3. However, this process does not contridict with genecolor principle. The positive and negative are the property of the bracket. They are attached on the F-colors. And the F-colors and gene-colors are independent from each other. When a pre-quark left bracket, it does not be controled by bracket's sign any more.

The b-quark decaying diagram is on the next page. However, it is not a complete picture. All branches of b-quark decay are not independent from one other. When one branch takes place, it forces another branch to begin. A lot of different kind of end products can be produced by the b-quark decay. Such as, $e^-$, $\mu^-$, $\tau^-$, $\bar{u}$, $\bar{c}$, etc. The genecolor force deter - mines the number of the brackets of the end products.

The F-color force determines the position (seats) of the Angultron.

When F-color force pulls all the -A together (form e⁻ ū, $\zeta^-$) from different colored b-quark, some Angultron are created because of the concept of pair-ocean. This universe is full of pair-prequark (A, -A). See next section.

VI. Neutron decay.

I have proposed a concept of pair-oceans. This

universe is full of pair particles. They are so losely bound, a huge energy is required to annihilate them. And, a u-quark can never turn into a d-quark. But (u - ū) quark pair is able to turn into a (d - d̄) quark pair and vice-versa. This can happen because of the concept of pair ocean. I call this pair transformation principle. And this process will change one pair quark's F-color. Neutron decay is a pure F-color decaying process.

"Why neutron is stable in nucleus, but it decays when it is free." In nucleus, neutron has the following structure and it is very stable.

$$
\text{neutron}
\left\{
\begin{array}{l}
u(\text{blue}) \left\{ \begin{array}{l} A \\ A \\ Vl \end{array} \right\} \text{Blue} \\[2em]
d(\text{anti-R}) \left\{ \begin{array}{l} -A \\ -V \\ -Vl \end{array} \right\} \\[2em]
d(\text{anti-Y}) \left\{ \begin{array}{l} -V \\ -A \\ -Vl \end{array} \right\}
\end{array}
\right\}
\left.
\begin{array}{l}
\\[2em]
\text{Anti-Blue}
\end{array}
\right\} \text{F-colorless}
$$

When neutron is free, it expose itself in the pair ocean. The blue u-quark will be attracted by an anti-blue (original) particle according to color force priority. Therefore, the following transformation will be taken place for neutron.

Isolated neutron = {u(blue), d(anti-R), d(anti-Y)}

--> free neutron = {u(B), d(Ba), d̄(B), d(Ya), d(Ra)}
According to pair transformation principle, it will transfer as,

----> {u(B), ū(Ba), u(Y), d(Ra), d(Ba)}
The final result is, n --> p e⁻$\nu_e$ . The free neutron decay mechanism is as follow.

$$u(B) \quad \left\{\begin{matrix} A \\ A \\ Vl \end{matrix}\right\} \rightarrow \left\{\begin{matrix} A \\ A \\ Vl \end{matrix}\right\} \longrightarrow \left\{\begin{matrix} A \\ A \\ Vl \end{matrix}\right\} \quad u(blue)$$

$$d(Ra) \quad \left\{\begin{matrix} -A \\ -V \\ -Vl \end{matrix}\right\} \rightarrow \left\{\begin{matrix} -A \\ -V \\ -Vl \end{matrix}\right\} \longrightarrow \left\{\begin{matrix} -A \\ -V \\ -Vl \end{matrix}\right\} \quad d(anti\text{-}R)$$

$$d(Ya) \quad \left\{\begin{matrix} -V \\ -A \\ -Vl \end{matrix}\right\} \rightarrow \left\{\begin{matrix} (-A) \\ -A \\ -Vl \end{matrix}\right\} \begin{matrix} \longrightarrow \\ \longrightarrow \\ \longrightarrow \end{matrix} \left.\begin{matrix} -A \\ -A \\ -Al \end{matrix}\right\} e^-(colorless)$$

$$d(Ba) \quad \left\{\begin{matrix} -V \\ -V \\ -Al \end{matrix}\right\} \rightarrow \left\{\begin{matrix} -V \\ -V \\ -Al \end{matrix}\right\} \begin{matrix} \longrightarrow \\ \longrightarrow \\ \longrightarrow \end{matrix} \left.\begin{matrix} -V \\ -V \\ -Vl \end{matrix}\right\} \nu_e(colorless)$$

$$\bar{d}(B) \quad \left\{\begin{matrix} V \\ V \\ Al \end{matrix}\right\} \rightarrow \left\{\begin{matrix} (A) \\ V \\ Al \end{matrix}\right\} \longrightarrow \left\{\begin{matrix} A \\ V \\ Al \end{matrix}\right\} \quad u(Yellow)$$

## VII. Color force and QED.

A) Free electron and tempo-state.

Free electron is F-colorless with a genecolor 1. It is a composite of three Angultron. Prequark does not possess colors (gene- or F-colors) by itself. However, it does have power to change F-particle's color. When an Angultron falls out from its seat, then a F-colorless e(+) will become a F-colored quark pair.

$$e(+) \quad \left\{\begin{matrix} A \longrightarrow A \\ A \quad \begin{matrix} V \\ \searrow A \end{matrix} \\ Al \searrow Al \end{matrix}\right\}_{u(R)} \left.\begin{matrix} A \\ V \\ V \end{matrix}\right\} (PaR\ \bar{x}) \quad \text{F-colorless}$$

Inside the dash-line, this is a tempo-state. The force for this reaction is called tempocolor force. For F-color, this tempo-state is stable. However the $\bar{x}$ (genecolorless d) is very unstable for the reason that genecolor conservation law is temporary violated. Therefore, the strong genecolor force pushes the tempo-state to go back to e(+).

 B) e(+), e(+) interaction.
 When two e(+) approach each other, two tempostates begin to interact.

$$
e(+)\left\{\begin{array}{ccc} A & \longleftrightarrow & A \\ A & \longleftarrow & V \\ Al & & V \end{array}\right\}(PaR)
\qquad
\begin{array}{c} V \\ V \quad (blue) \\ (Al)\longleftarrow \end{array}
$$

$$
\left.\begin{array}{c} V \\ \hookrightarrow A \\ \longrightarrow Al \end{array}\right\}(R)
\qquad
(PaB)\left\{\begin{array}{ccc} (A) & \longleftrightarrow & (A) \\ (A) & \longleftrightarrow & (A) \\ V & \hookrightarrow & (Al) \end{array}\right\}\ e(+)
$$

Now, PaR has two choices to utilize its genecolor force. When PaR, blue interaction has taken place, then these two tempostates become one composited F-colorless tempostate.

$$
PaR\left\{\begin{array}{ccc} A & \xrightarrow{\phantom{xx}} & A \ (Y) \\ V & & V \\ V & & (Al) \end{array}\right.
\quad
\left.\begin{array}{c} V \\ V \\ (Al) \end{array}\right\}(B)
$$

$$
e(+)
\qquad\qquad\qquad\qquad\qquad\qquad e(+)
$$

$$
Red\left\{\begin{array}{ccc} V & \longrightarrow & V \\ A & \longrightarrow & A \\ Al & \longrightarrow & Al \end{array}\right\}
\qquad
\left\{\begin{array}{ccc} (A) & \longleftarrow & (A) \\ (A) & \longleftarrow & (A) \\ V & \longleftarrow & V \end{array}\right\}PaB
$$

Inside the dash-line, there are three u-quarks. They are F-colorless. However, PaB is very unstable because genecolor conservation law is violated. Therefore, this composited tempostate will move into a stable state as follow. For a free e(+), it went

into a tempostate and returned back to itself. However, for a two e(+) system, a photon is automatically
created to mediated the interaction between them.
Vacutron and bracket is not conserved.

C) e(+), e(-) interaction.

This interaction is similar with e(+), e(+) interaction with a small difference. The two steps is as
follow.

STEP 1.

The tempostate inside the dash-line is slightly different from e(+), e(+) interaction. A mixed yellow
quark is created. This quark has zero charge with
a genecolor 1. In e(+), e(-) interaction, there is no
Angultron exchange. Therefore, it is an attractive

force.
STEP 2.

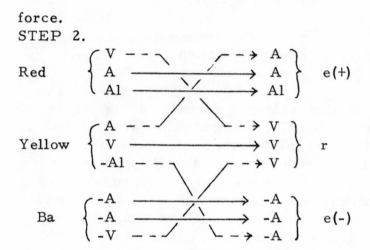

VIII. Conclusion.

QED assumption has been derived by using the concept of color force.    Some particles' decay have also been explained by color forces.    In this paper, I have defined 7 colors --- 3 genecolors, 3 F-colors and one colorless (gene- & F-color).

References:
Greenberg O. W.  1978 "Quarks".  The annual review of Nuclear & particle physics.
Appelquist T.  1978 "Charm & Beyond".  The annual review of Nuclear & particle physics.

# CHAPTER THREE

## Super Unification

Why is this universe triplicated? What is graviton? What is charge? What are colors? In order to answer all these questions, we have to redefine what is physics. We are always trying to describe nature. We are always waiting for the experiments to verify the theory. Now, we are ready to explain nature. In this paper, I have discussed all questions mentioned above. And I have quantized gravity and geomentized quanta at the same time.

## I. Introduction.

The quantum electrodynamics, quantum chromodynamics and electro-weak theory are most successful physical theories ever developed. Most of their predictions (except of a few) have been verified by the experiments. Here, I will briefly list the major features and the major dilemmas of standard model.

A) The major features.

1. For the known properties of matter, 24 particle (quarks and leptons) are needed to represent them all.

2. All observed interactions (except gravity) of matter can be explained as exchanges of 12 gauge bosons. All of them are massless except of three weak bosons. Some of them are colorless but some are not.

3. As a fundamental particles, leptons (integral charge) are colorless, and quarks (fractional charge) are colored.

4. QED, QCD, electroweak theory are discribed

24

by essentially the same mathematical structure ----
gauge-invariant field theory. They have to be renormalizable. All forces are transmitted from one
particle to another by carrier fields.

5. QED, QCD, electroweak theory are all symmetrical theory. However, the weak force results a
spontaneous symmetry breaking. And the weak charge
is not conserved.

6. Everything is triplicated.

B) The major dilemmas.

Every major feature discussed in standard model
is also a major dilemma for itself. Also, we have
the following difficulities.

1. Why are 3, 6, 12, 24 repeated in nature over
and over? 3 colors, 3 generations, 6 leptons, 6 quarks,
12 gauge bosons, etc.

2. Is 4/3 unit of charge existing?

3. The grand unification is not nearly in sight if we
have to introduce new force field one after another.

4. There is not any hope to unify the gravity with
any of old and new force fields (QED, weak, color,
hypercolor, etc.) in standard model and its success
theories.

In this chapter, I will try to discuss all the topics
mentioned above.

II. Three assumptions.

The following three assumptions can outline a
picture for a super unified field theory.

A) Assumption I.

I define charge as follows, $q = \pm \sqrt{L \cdot c}$, L is the
angular momentum around the time coordinate. L is
a spin-like but it is not particle's spin. Actually, L
is the result of a quantized space which has $\Delta\theta$ with
absolute time coordinate. (see chapter four).
When $\Delta\theta = \pi/4$, $L = h/2$, $q = \pm \sqrt{L \cdot c}$ (unit charge).

$\Delta\theta_1 = 13.43^o$, $L = h/18$, $q_1 = q/3$

$\Delta\theta_2 = 28.80^o$, $L = 4h/18$, $q_2 = 2q/3$

Coulomb force is the statistics sum of the interaction between the charge and time-space. Let $q_1 = n\,q$, $q_2 = m\,q$, then,

$$F(\text{Coulomb}) = (mn/137)(hc/r^2) = f_1\,hc/r^2 \ldots \text{(1)}$$

$f_1$ (coupling factor) is dimensionless.

B) Assumption II.

I define the gravitational constant (G) as,

$G = hc/a^2$          a has dimension as (kg).

Then, we can rewrite Newton's law as,

$$F(\text{Newton}) = G\,m_1\,m_2/r^2 = (hc/a^2)(m_1\,m_2/r^2)$$

$$= (m_1\,m_2/a^2)(hc/r^2) = f_2\,hc/r^2 \ldots\ldots\ldots \text{(2)}$$

C) Assumption III.

I define space as follow, $\Delta s = (i^{n1}, i^{n2}, i^{n3})\,c\,\Delta t$, and $n1$, $n2$, $n3$ are 1, 2, 3, 4. So far, there is absolutely no physical evidence to suggest that space or time itself is made of little quantal pieces. However, I am going to assume that time is a quanta. I define time as follow.

1.  $\Delta t$ is a quanta, $\Delta t = h/mc^2 > 0$.

2. For our convenience, we define time from our measuring.

3. To accurately observe and discuss a relative motion, we have to coincide the time of two watches in the system.

Now, space and time are clearly defined. These definitions are completely different from the conventional. Assumption III is the most important concept in this unified theory. Parity, colors and graviton are the direct consequences of assumption III.

$$N^2 = (i^{n1}, i^{n2}, i^{n3})^2 = \left\{\pm 1, \pm 3\right\} \qquad \ldots \text{(3)}$$

When a subspace has $N^2 = \pm 3$, then this subspace

is a vacuum.    When there is a relative velocity (v),
assumption III becomes,

$$N^2 \, c^2 \Delta t^2 = \Delta s^2 + b \ldots (4)$$

III. Source of force fields.

We always define mass with operational defination.
However, I will define mass with a different approach,
in part based on theological principles.   But, after we
UNIFY and GENERALIZE this new defination,   the
equation 9 will become operational verifiable again.
I will discuss this in two steps.

First, we have noticed the difference between time
and space.   Therefore, I have to define two different
kinds of mass here.

1) $m_t = h / \Delta t \, c^2$ . . . . . . time-like mass.

2) $m_s = h / \Delta s \, c$ . . . . . space-like mass.

The relationship of $m_t$ and $m_s$ is as, $m_t = N \, m_s$.
When two particles interact, a force is between them.
And this force can be derived as follow.

Let $m_1 = m_{s1} \, N_1$,   $m_2 = m_{s2} \, N_2$   and   $m_1 = m_2$

So,  $M_t = m_1 \, m_2 \, / (m_1 + m_2)$

$$F_{12} = c \, \partial M_t / \partial t = ( h / \Delta t \Delta s)(1/2)(N_1 N_2 / N_1 + N_2) \ldots (5)$$

Here, $\Delta s = h / m_s c$,  $\Delta t$ is the time period of the
interaction.   And,

a) The longer $\Delta t$ is, the  weaker F is.

b) The large of mass  $m_s$ is, the stronger F is.

c)  K $= (1/2) (N_1 N_2 / N_1 + N_2)$ is the major factor for
determing the magnitude of F.

From assumption III and Eq(3), we can derive K as,

$$K = \sum_{n=0}^{\infty} K(n) = \sum_{n=0}^{\infty} (-1)^n a_n / 64^n \ldots (6)$$

When n = 0, it is strong interaction.   When n = 1, it
is Coulomb force. When n = 7, it is weak interaction.

When n = 22, it is gravitational force.

Second, even if there is a difference between time and space, assumption III unified them. Now, we could find its unified meaning. I can redefine mass as,

$$m = (m_s \, m_t)^{1/2} = (h/c^3)^{1/2}(h/\Delta t \, \Delta s)^{1/2}$$

$$= (h/c^3)^{1/2} (F)^{1/2} \quad \dots\dots\dots\dots\dots \quad (7)$$

Eq(7) has telling us that mass is the source of all force fields. Therefore, mass should contains all kinds of informations (such as, parity, colors, etc.). Actually, it does. It does become OPERATIONAL VERIFIABLE. We can rewrite Eq(7) as,

$$m = (h/c^3)^{1/2} \sum_{n=0}^{\infty} (i^b)^n (1 - v^2/c^2)^{1/2}(a_n/64^n)^{1/2}$$

$$x \, (h/\Delta t \, \Delta s)^{1/2} \quad \dots\dots\dots\dots \quad (8)$$

$$m = (h/c^2 \Delta t) \, (1/N)^{1/2}(1 - v^2/c^2)^{1/2}$$

$$x \sum_{n=0}^{\infty} (i^b)^n (a_n/64^n)^{1/2} \quad \dots\dots\dots \quad (9)$$

a) In Eq(9), $C = (1/N)^{1/2}$. This term is associated with charges, F-colors and genecolors (generations). (see later chapters).

b) $m = h/\Delta t \cdot c^2$ is the unit and the amount of particle's mass.

c) $r = (1 - v^2/c^2)^{1/2}$ is the relativity effect.

d) $\sum_{n=0}^{\infty} (a_n/64^n)^{1/2}$ is the coupling constant for the particle. $(1/64^n)$ is the major coupling factor. $a_n$ is the minor coupling factor. In QCD, electroweak theory, we have discussed $a_n$ in detail.

e) Let me make another suggestion. I define parity as, $p = (i^b)^n$. b is a very special quantum number. It can be 1, 2, 3 or 4 only. It is the result of the time

Coordinate which has degenerated to 4 subcoordinates.
If a particle has b = 2, then it will decay to 2 particles.
If an interaction has $b = b_1 * b_2 = 3$, then it will produce a 3 jets event. Now, we can check some particles' parity.

$$p = ( i^1 )^0 = +1, \qquad\qquad n = ( i^3)^0 = +1$$

$$\pi = ( i^2)^1 = -1, \qquad \mu = ( i^3)^7 = +i \text{ or } \mu = ( i^3)^1 = -i$$

$$e = ( i^1)^1 = +i \quad \text{or} \qquad e = ( i^1)^7 = -i$$

Very clearly, parity is not conserved for lepton on
the conventional sense.

IV. A new uncertainty principle can be derived.
$\Delta E = \Delta W = F \Delta s = K ( h/\Delta t \Delta s) \Delta s$
So, $\Delta E \Delta t = K h$.
And, $\Delta F \Delta t = \Delta p = K ( h/\Delta t \Delta s) \Delta t$
So, $\Delta p \Delta s = K h$.
Heisenberg's uncertainty principle is not a fundamental law of nature, but depends on the whole theory
of quantum mechanics. However, this new uncertainty principle is a fundamental law of nature. The
effectiveness of this principle depends on the coupling constant K. Very obviously, this principle is not
very effective for the gravitational interaction because
that K is just too small ( $K \sim 10^{-40}$). The following
is a diagram of how UF degenerates to four forces
according to NUP and how the K scale forms a loop
because of the asymptotic freedom. And it represents
two states of this universe -- degenerated (during and
after big bang) and non-degenerated (before big bang).
(see chapter six).
UF -- Unified force.  NUP -- new uncertainty principle.
QM -- quantum mechanics.  CM -- classical mechanics.
EWT -- electroweak theory.

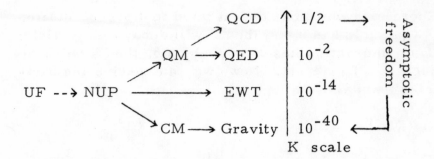

V. The structure of nucleus is a double-spiral.

I am going to describe this by using the magic number.

1) According to selection rule Eq(3), only following subspaces contain matter. Example: When $n_1$ = 4, $n_2$ = 1 or 3 then $n_3$ can be 1, 2, 3 or 4. So, total livable subspaces are 1 x 2 x 4 = 8. And the basic livable structures (contain matter) are 2, 6, 8. The principle of superposition is true in this theory. So, 2, 4, 6, 8, 12, 16, 22 = (16 + 6), 32, 44 are suppose to be the livable structures. The relationship between the magic numbers and livable structures are as follow.

| Magic # | 2 | 8 | 20 | 28 | 50 | 82 | 126 |
|---|---|---|---|---|---|---|---|
| Liv. Stru. | 6 | 12 | 8 | 22 | 32 | 44 | |

2) Space structure and magic numbers.

i) The space which the protons (neutrons) occupied is $\triangle$ s.

ii) The space which a shell has is the circumference, not area or volume. This is a one dimensional universe.

Let,

A is the max. number of usable space (integer) of $n\pi$.
B is the max. number of protons is allowed by livable structures.

$C_1$ is the sum of A (total usable spaces).

$C_2$ is the max. integer number of $\sum_{1}^{n} n\pi$

D is the magic numbers.

| n | Dia. | Circum. | A | B | $C_1$ | $C_2$ | D |
|---|------|---------|----|----|-----|-----|--------|
| 1 | Δs | πΔs | 3 | 2 | 3 | 3 | 2 |
| 2 | 3Δs | 3πΔs | 9 | 6 | 12 | 12 | 8 |
| 3 | 5Δs | 5πΔs | 15 | 12 | 27 | 28 | 20, 28 |
| 4 | 7Δs | 7πΔs | -- | -- | -- | 50 | ⎫ 50 |
| 5 | 9Δs | 9πΔs | 28 | 22 | 55 | -- | ⎭ |
| 6 | 11Δs | 11πΔs | 34 | 32 | 89 | -- | 82 |
| 7 | 13Δs | 13πΔs | 40 | (44) | 129 | -- | 126 |

# CHAPTER FOUR

## The geometry of the unified universe

Special relativity theory is the direct result of a topological property of the unified space which is independent from any influence of observers. Furthermore, the way our nature is trisecting an angle has created charge.

I. Introduction.

Are we able to search physics which is beyond the Planck's time scale. We might run into some difficulty on the experimental aspect. Maybe we have to change our philosophy.

Newton assumed that our nature space is an Euclidean space. Einstein argued that our nature has Riemannian space. We all know the difference between geometry and physics. Physics discuss the "dynamics" of a group particles which possess mass. Geometry discuss the "statics relationship" of a group points which are massless. Seventy years ago, Einstein intended to unify physics and geometry. He tried to geomentize the gravity. However, one philosophy concept of his has blinded his eyes. He put too much emphasis on "the observer" during any kind of transformation. H also very emotionly tried to maintain the gauge and symmetry invariance in all kinds of transformation. Very obviously, there are some very simple transformations existing in nature which do not have gauge and symmetry invariance. The simplest example is to transfer a circle to a square which has the same area. For a circle, the degrees of symmetry to the origin is infinitive. I defined, "the degree of symmetry" is the numbers of transformation (example, rotation) which can be made under symmetry invariance condition. However, after the transformation, the degrees of symmetry is four for the square. This transformation (circle --→ square) is actually to transfer "infinity" to "finite". Here, I am going to use two simple examples to discuss some characteristics of "the nature space". Such as, rigidness, softness, flatness, infinity, and its statical or dynamical properties.

II. Trisect an angle with a compass and straight edge.

It is very easy to trisect any angle with a compass and straight edge only by using the following procedure. (see Fig. 1, on page 39).

i) Arbitrarily choose a real number r as a radius. Use the vertex (point 0) of the known angle $\alpha$ as the origin, r as the radius and draw a circle which intersects with angle $\alpha$ at point A & B. This circle also intersects line OA at Dl. Give this circle a name as circle (0).

ii) Use Dl as origin and r as radius draw a circle which intersects line OA at Cl.

iii) Connect Cl B with a straight edge. And line Cl B intersects circle (0) at point D2.

iv) Use D2 as origin and r as radius draw a circle which intersects line OA at C2.

v) Connect C2 B with a straight edge. And line C2 B intersects circle (0) at D3.

vi) Continue the above procedure. Use D(n) as origin and r as radius to find C(n). Then C(n) B intersects circle (0) at D(n + 1).

vii) When n --→ $\infty$,
then angle B $C_\infty$ A = ( 1/3) Angle $\alpha$
From the above discussion I obtain the following con-

clusions.

1) We are able to trisect any angle with compass and straight edge even if it takes forever.

2) We can not trisect an angle with a compass and straight edge by any finite number of steps.

3) Any space (curved, etc.) can be approximated by using the Euclidean space (rigid and flat). For example, a 4 dimensional curved space can be represented by a 10 dimensional Euclidean space. This is why gauge field theories have been so successful in the past. And it will continue to do so in the future. This is also why we have a flatness problem on the cosmology. Our universe is nearly flat, and it can be represented by a flat space to any desired accuracy.

4) The above procedure has proved that "the infinity" is the basical characteristics of our universe. It is the link between "nature" to a "flat space". (see chapter seven and eight).

III. Transfer a circle to a square which has same area.

I have successfully used a compass (rigid) and straight edge (flat) to trisect an angle by introducing the concept of "the countable infinity". However, the same tactics is useless for transfering a circle to a square. This result hints that "the nature space" at least has one additional characteristics besides "the countable infinity". In other wards, "THE COUNTABLE INFINITY" I introduced above is not able to reach this HIDDEN PROPERTY. Now, let's re-examine our problem. To transfer a circle to a square (same area) is the same as for trying to find $\sqrt{\pi}$ by using an arbitrary real number r (radius). When we choose $r = 1$, then the task becomes an algebra problem. How to find $\sqrt{\pi}$ ? Very obviously, the compass and straight edge are out of the ball

game.  In chapter  VII,  $\pi$ is "+" colored.  Furthermore  it  possesses  a  greater , stronger  power  of
"infinity"  than  the one I introduced above (the countable infinity).   But, what is "the infinity" which has a
stronger power than the "countable infinity"?

If we can find a method to tranfer circle to square
(same area),   then  we  should understand what is the
necessary  ingrediant  for this transformation.  Let's
investigate the following procedure. (see Fig. 2).

i) Let  point  A  as  origin,  and use a SOFT compass
(able  to  stretch  the  compass open uniformly) draw a
curve  when  the  compass  is  UNIFORMLY stretched
open.

ii) After  a  full turn ( $0 \rightarrow 2\pi$ ), it reached point B.
Then make a tangent line BC at point B.

iii) Connect  line  AB, and make a line AC perpendi-
cular  to  line  AB and intersects with BC at point C.
Very obviously, the area of circle A (radius =line AB)
is equal to the area of triangle ABC.

iv) Transfer triangle ABC to a square (same area).

I  have  successfully  transfered  a  circle (radius =
line AB) to a square with a soft compass which moves
outward  uniformly.   Therefore,  $\pi$  possesses  two
characteristics --- softness and dynamical property.
A portion of our "nature space" can not be reached by
a  rigid  gauge  because  it is soft.  (se chapter VIII).
Also, the transformation  from  "infinity" to "finite"
has to ride on a dynamical system.  In a rigid space,
the  gauge  and  symmetry  invariance  are  normally
preserved.   However,  in  a  soft  space, gauge and
symmetry  invariance  are  no  longer  the necessary
properties  of the space.  The nature space has per-
fectly  topological  invariance  except  under  " one
transformation".  (see chapter VI).

IV. Creation of charge and mixing angles.

From assumption III, 1 time dimension degenerated to 3 space dimensions. Every space dimension degenerated to 4 time subcoordinates, then it further degenerates to 64 subspaces. When a particle is created, it will immediately possess a space-time coordinate in this degenerated space-time space. One most important component of this space-time coordinate is an angle which is also quantized. And charge is the direct result of this quantized space-time ANGLE. The following equations are how an angle was trisected in a curved space.

i) $\sin \Delta \theta_1 = \sin^2 \Delta \theta_2$

ii) $(2 - 1/6) \theta_b + \Delta \theta_1 + \Delta \theta_2 = \pi/4$

iii) $\theta_b = (1/2) \sum_{n=1}^{\infty} (\pi/64)^n = 1.47895^\circ$ (bending angle)

Therefore, $\Delta \theta_1$ --- $13.43^\circ$ (Cabibbo angle)

$\Delta \theta_2$ --- $28.80^\circ$ (Weinberg angle)

And, $\sin \Delta \theta_2 \sim 2 \sin \Delta \theta_1 \sim (2/3) \sin(\pi/4)$
Very clearly, from the defination of charge, $q = \pm \sqrt{L \cdot c}$

$\Delta \theta = \pi/4$, then q is one unit of charge.

$\theta_c = \Delta \theta_1 = (1/24)(2\pi - 24 \theta_b)$, then $q_1 = q/3$

$\theta_w = \Delta \theta_2 = (1/24) ((2\pi - 2 \theta_1) + (2\pi - 2 \theta_b))$

The bending angle $\theta_b$ is the only independent variable for quantizing the $\Delta \theta$. Cabibbo and Weinberg angles are the most important candidates for unifying three forces in the standard model. I have derived them in the above discussion. However, some other angles also can be found. A generalized principle is as follow.

$$\sin \Delta\theta_1 = \sin^2 \Delta\theta_2 = ( \sin^2 \Delta\theta_3)^2 = (\sin^3 \Delta\theta_4)^3$$

$$= (\sin^6 \Delta\theta_5)^6 = (\sin^{64} \Delta\theta_6)^{64} \ldots\ldots (10)$$

Even with this generalized principle, the nature is not able to divide space prefectly even. Nature can not trisect an angle with finite number of steps.

Cabibbo and Weinberg angles are both free parametters in their theories. I have given them some new physical meaning in the above discussion.

## V. Conclusion.

Our "nature space" is soft, curved and dynamically driven to "the infinity". However, it can be very accurately approximated with a rigid, flat and statical space (such as Euclidean space). It means that "the observer" is not very important to our nature. The softness and rigidness are both intrinsically existing in the nature without any influence from any observers. However, the concept of "observer" is very useful (or necessary) for us (human) to understand nature. We (observers) are able to observe the nature's softness and rigidness. However, the nature's characteristics of "the infinity" is unobservable . (see chapter ten). We can only observe the results (colors, charge etc. ) of those "infinity". The "infinity" is the only cause for a rigid space to become soft. We are not only able to compress the time and rigid rod (Einstein theory) but also are able to compress all numbers of 1 to $+\infty$ into a small interval $[0, 1]$ . Actually, $[0, 1]$ and $[1, \infty]$ are isomorphic. Farthermore, we can squeeze $[1, \infty )$ (metric space) into an infinitesimal interval $B = [0, 1/10^n]$ , $n \in N$. However, B is very rigid. There is no way to compress B to Wu $= [0, 0]$ without lose the topological invariance. When $n = \infty$ ,

there is "one transformation" which is not topological
invariant. This "one transformation" which is not
topological invaried created our universe. (see chapter
VI). After the birth of our universe, every transformations have to be topological invariant. However,
this topological space can be very accurately approximated by a rigid, flat space which possesses very
good gauge and symmetry invariance.

Our universe is a one dimensional space --absolute
time. Time quanta's energy goes into matter and
space (see chapter V). The trajectary of time quanta
is a time cone. Its projected view is an Archimedes'
spiral. This spiral is the only tunnel (or funnel) to
transfer "infinity" to "finite". Amazingly, it also
trisected the singular pole at the same time. (In Fig. 2,
the area under line AB and curve AB is 1/3 of circle A
(radius = AB) ). The singular pole is the source for
colors (see chapter V). Now, physics has to deal with
and accept the concept of "the infinity" which can
never derived by any possible experiment. Is the
interplay between experiment and theory the only
correct way for understanding of the law of nature?
Now is the time for us to re-examine our philosophy.

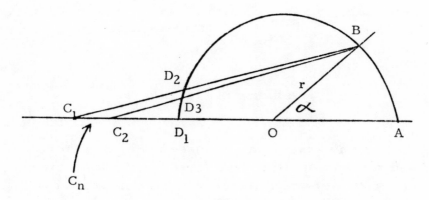

FIG. 1) We are able to trisect any angle with compass and straight edge with countable steps.

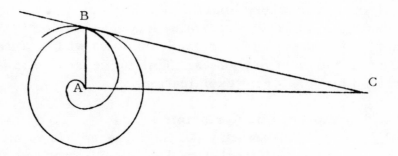

FIG. 2) A portion of our "nature space" can not be reached by a rigid gauge because it is soft. Archimedes' spiral is the only tunnel to transfer "infinity" to "finite".

39

# CHAPTER FIVE

## The future of the universe

Is our universe open or closed? Why has there been no anti-matter star observed? Does the energy conservation law have to be violated for a closed universe? According to my SUT, this universe is open if and only if it has bounced infinite number of times. Furthermore, with a delayed supersymmetry principle, the matter and anti-matter alternately appear in each "Big Bang". Also, the energy conservation law is obeyed for a closed universe.

I. Introduction.

Today, the most evidence and opinions are favoring an open universe. Such as, the observed (visible matter) and calculated (dark matter) density of this universe is barely making up 50% of the critical density. I am very certain that the contribution of neutrino mass for a closed universe will be ruled out in the next few years. This universe has to be closed by a different explaination.

II. The super unified principles.
 i) Two poles structure --- this universe has to move from pole 1 (big bang) to pole 2 (big crunch or the infinity?). The concept of two poles structure actually came from the research of elementary particles. All fundamental particles (quarks and leptons) can be

clearly identified with three and only three independent quantum number --- charge (quantized angle), F-color and genecolor. These three quantum number form a topological structure which is a torus. This torus has two poles, one outer surface and one inner surface. The pole 1 is trisected by Archimedes' spiral (see chapter IV) and they are genecolors (1, 2, 3). The pole 2 is also trisected and they are F-colors (Red, Yellow, Blue). The outer surface is topological equivalent to a cylinder surface which is embedded in the torus. Therefore, the parity for leptons are not conserved and they are F-colorless and possess integer charge. But, the quark trajectories are on a Mobius band which is also embedded in the torus.

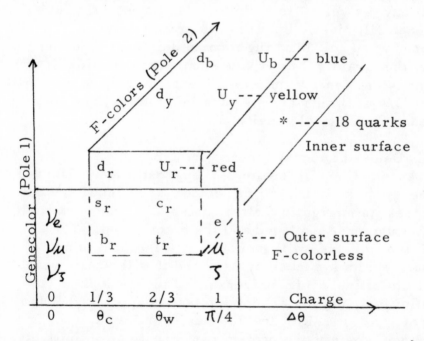

ii) Assumption III --- $\Delta s = N c \Delta t$ & $N^2 = \left\{ \pm 1, \pm 3 \right\}$ Time is created from nothing (see chapter VI). Every time quanta possesses mass (energy). This time

quanta's energy will transfer into two forms.   One is
matter form when $N^2 = \pm 1$.   The other is space
(vacuum) form when $N^2 = \pm 3$.   During the expansion
period, the  space of the universe increased when the
time goes  forward.   1/4 of the time quanta's energy
goes into space (vacuum) which acts as an energy
bank.   During the contraction period, this  saved
energy will be released which cause a psedo energy
creation phenomenon.   The energy conservation law
seems to be violated during this period.

   The real meaning of assumption III is that, all mass
(energy) of the universe is coming from the time.
Actually, an open universe is prohibited by assumption
III.  The big bang momentum is finite unless the mass
of our universe· is infinite.   And, the total gravita -
tional will diverage unless the gravity is zero.  Therefore, when the big bang momentum was exhausted out
by the gravity, the space expansion will slow down.
And, the density of this universe will increase slow-
ly (3/4 of time's energy goes into matter) and  finally
reach the critical density.

III.  Delayed supersymmetry principle.
   Assumption III seems to  suggest a nonconserved
energy principle.    Time comes from nowhere and
stores its energy in matter (mass) and space (vacuum).
This universe exploded and them collapsed.   The size
of the universe will increase by a factor 2 (approximately) during  each cycle.   And it will finally reach
an infinitively big universe.   The concept of "the
infinitively big" is acceptable (see chapter VII).   However,  it is unimaginable that there is an energy
source somewhere which can  create an infinitively
big universe.

   Today, all of the physicists believe that anti-matter
is part of God's creation.   But why have not been able

to observe an anti-matter star.    There are two reasons.

i) Pole 1 (big bang) and pole 2 (big crunch) are different.  Pole 2 is a second big bang for another cycle. If the 1st cycle universe is constructed with anti-matter, then 2nd cycle universe will be constructed with matter.  (see Fig. 3, on page 45).    Also, the symmetry is delayed and not perfect around the absolute time.

ii) Energy conservation law ---- if we arbitrarily define that anti-matter possesses positive energy and matter possesses negative energy.  And let A1, A2, A3, ... represent the total energy of cycle 1, 2, 3 ... Then the total energy is conserved.

$$\sum_{n=1}^{\infty} A(n) = 0$$

IV. Genecolors.

When the universe becomes infinitively big, the meaning of open and close lost their meaning.  It is closed because the expanded space will intersect with absolute time at an infinite far away place (see Fig. 4), and then a new big bang begins.  However, it is also open because the space will expand forever before it begins to contract.  All predicted phenomena for an open universe will be realized, such as,  Proton decay, collapsed Galaxy - black hole, and evaporating black hole, etc.  In Fig. 4, the life time of universe $A_{\infty}$ is infinitively long.  Let's give this "infinity" a name "inf P".  However, the universe $A_{\infty}$ will intersect with absolute time at the point $\infty + 1$, and a new universe $A_{\infty+1}$ begins.  The number cycles of universe are $\infty^{\infty^{..}}$.  Now there are three "infinity". (see Fig. 4)

a) $I = \infty$ (countable), number cycles.
b) II = inf P.
c) III = Line(1, 2) + Line(2, 3) + ... + Line(I, I+1) + ...

Obviously, inf P is not physically countable. If inf P has highest power of an infinity, then the existance of a countable infinity becomes unnecessary in our universe. If inf P has a power less than the highest power of infinity such as --- eternity, then an infinitively big universe will intersect with TIME which has higher infinitive power. Therefore, three kinds of infinity are necessary. In elementary particles, the properties of these three infinities were showed up as genecolors. (see chapter VII & VIII).

V. Conclusion.

There are only three assumptions introduced in my SUT. And a lot of physical phenomena have been explained. Assumption I & II can be very easily verified by the physics means. Assumption III is coming from the faith for the "God" who is super-symmetry at the point of inf U (the eternity). When a big bang takes place, the expansion can not be visualized from the outside by any observer who possesses mass. There is no external vantage because what is exploding is the entire universe. This new born universe is the direct consequence of a previous one. Furthermore, it is the direct cause for the next big bang.

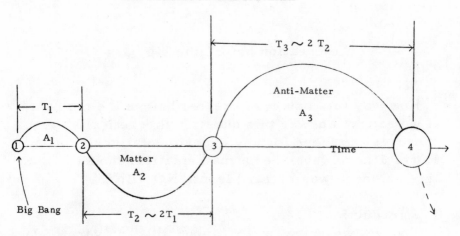

FIG. 3) Matter and anti-matter alternately appear in each "Big Bang". And the size of the universe will increase by a factor 2 during each cycle.

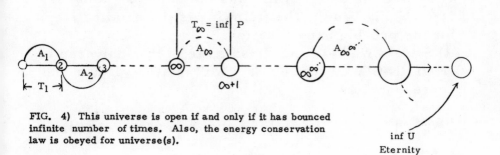

FIG. 4) This universe is open if and only if it has bounced infinite number of times. Also, the energy conservation law is obeyed for universe(s).

45

# CHAPTER SIX

## The creation before the big bang

How was this universe created before the big bang? If we can not answer this question theoretically, then there is no hope for any experimental verification. Here, I have developed a mathematical model of how this universe was created from "NOTHING".

I. Introduction.

Today, we know a lot about this universe. In particle physics, there are 24 fundamental particles and 12 gauge bosons. In astro-physics, we know very much the detailed evolution of this universe. However, for any gauge theories, there is no way to address how this universe was created before the big bang.

II. The creation before the big bang.

This universe has evolved for billions of years after the big bang. What was the universe doing before the big bang? The following is the detailed steps for the pro-big bang creation.

1) State "Wu" --- the first state of this universe is "NOTHING". I called it "Wu" state. It can be represented as,

$$Wu = \begin{bmatrix} 0, & 0 \end{bmatrix} = \begin{bmatrix} 0, & 1/10^n \end{bmatrix} \text{ and } n = \infty \quad \dots \dots (11)$$

If we can find a transformation to transfer " $\infty$ " to finite in a "natural way", then "SOMETHING" is created. I did suggest three way for infinity-finite

transformation.  One is by trisecting an angle.  One
is by using symmetry principle (see this chapter).
The other is by using the Archimede's spiral dynamics.
(see chapter IV).  Actually, these three methods
indicate three kinds of infinity. (see chapter VII).
  2) State "Yu" --- Yu means "something".  With a
symmetry creation operation, we can very easily
transfer "Wu" to "Yu".

$$Wu = (0, 1/10^m) = A1 + A2$$

$$= (0, -1/10^n) + (0, +1/10^n) \ \ldots\ldots\ldots (12)$$

m is infinite and n is finite.  Al, A2 are in the state
"Yu".   A2  is an infinitesimal interval which is topo-
logically equivalent (isomorphism) to $[1, \infty]$ .  Now
the creation procedure before big bang is very clear.
  i) Nothing, "Wu" --> Something, "Yu" (infinitesimal).
By a supersymmetry creation operation.
  ii) Finite $[0, n]$ --> Infinity $[n, \infty]$ , by a topological
transformation.  After finite size universe has bounced
infinitive times, universe becomes infinitive big.  To
create an infinitive big universe does not violate the
energy conservation law for the reason of supersymmetry.  The matter and anti-matter alternately
appear in each "big bang".
  3) Unified thoplogy (Unilogy)
  If we can find a reason to force the newly created
universe (Yu) to bounce, then we have completed our
logic.  In topology (combinatorial), there are many
different kinds of surface and space.  Such as, sphere,
torus, projective plane, etc.  If our universe is
unified, then all these surfaces have to also come
from the same origin.  If "Wu" is the "ORIGIN" for
everything, then these surfaces have to come from
"Wu" too.  Actually, it does.

$$Wu = ( \ 0 \ ) = A - B \ \ldots\ldots ( 13 )$$

A is Euler characteristic, B is the total number of black points (see chapter nine) of a surface. Eq(13) is the fundamental principle for unilogy which is discussed in chapter nine. According to Unilogy. creation begins from one black point (singular pole) and has to reach another black point. Therefore, the universe has to "bounce".

III. The big bang and gravity.

The special relativity has three fundamental assumptions.

i) All inertial frames are completely equivalent.

ii) The velocity of light is independent of the motion of its source.

iii) Time, space have operational definitions only. The third assumption implies that the absolute coordinate is not existing. But, the Michelson-Morley experiment is only proving that an at rest ether frame is not existing. Einstein's theory also has three well known and proved equations. The following is one of them.

$$m(v) = m(0)/ \ ( \ 1 - v^2/c^2)^{1/2} \ \ldots\ldots ( 14 )$$

These equations very clearly show everybody that there is not any particle which has non-zero rest mass $(m(0))$ can have a relative velocity equal to light speed. If a theory is trying to discuss a particle which has a momentum $p = mc$ and $m \neq 0$, then this theory has seriously violated the Einstein's theory --- the major corner stones for our modern physics. However, we can rewrite Eq(14) as,

$$m(0) c = m(v)( \ c^2 - v^2 \ )^{1/2}$$

A rest particle has an absolute momentum $p = m(0)c$, and if $v$(relative velocity) $= c$, then $m(0)$ has to be zero.

During the big bang, Eq(9) was reduced to a simple form.

$$m = ( h/c^2 \Delta t)( 1 - v^2/c^2)^{1/2}$$

This universe was a super hot and dense pile. There was no strong force because of the asymptotic free-dom. There was no weak and electromagnetic force because that the mixing angles has not yet been de-generated. The gravity is the only existing funda-mental force at that moment. And, it came from the absolute momentum $m(0)c$, $m(0) \neq 0$, and a relative velocity (v). It begins with Einstein's equation.

$$m(v) = m(0)/( 1 - v^2/c^2)^{1/2}$$

$$\rightarrow m(v) v = m(0)c/(c^2/v^2 - 1 )^{1/2} = \alpha \, m(0)c$$

And,

$$F(gravity) = m(v)v/\Delta t = \alpha m(0)c/\Delta t = \ldots = Gm_1 m_2/r^2$$

The unified force is $F = K \, h/\Delta t \Delta s$. And, the gravi-ton is the time quanta which has a mass m, $\Delta t = h/mc^2$.

However, if gravity was the only force during the big bang state, then the big bang could never happen. At big bang state, our universe was in an infinitive phase. The density and temperature were both in-finitively high. Therefore, there were abundant thermal photon and super-heavy neutral particles. At a millisecond right before the big bang, the ex-panding force caused by particles collision has to be in an equilibrium state with gravity force. Obviously, it finally had to overcome the gravity and the expan-sion began. For infinitive density, the density fluctuation is impossible and meaningless. However, as soon as the density of universe become finite, the space began to degenerate and mixing angles were created. Therefore, a fluctuation for density is inevitable, and galaxies were formed. The expanding force will decrease rapidly because of expanded sapce

and reduced temperature, and very quickly reached
to zero.   If the decreasing of gravity force is smaller
than the decreasing of the expanding force, then this
universe will begin to contract soon after big bang.
If the expanding force is infinitively large than gravity,
then the event of a big bang becomes impossible .
Therefore, gravity has to decrease a lot faster than
expanding force, and it takes billions more years to
catch up after the expanding force had ceased billions
of years ago.

THE GRAVITATIONAL CONSTANT (G) HAS TO
DECREASE RAPIDLY.   This is the direct result of
how the nature measure its time and space.   We can
define time unit (TU) as follows,

TU = TU(big bang) x   r(radius of big bang)/R(radius
of universe)

And, TU = $(G h/c^5)^{1/2}$.   The TU has to be larger than
$10^{-10}$ second at big bang state, otherwise the strong
and weak interaction will take place.   At big bang
state, time is clearly quantized. Today, TU$\sim 10^{-44}$.
Therefore, we can estimate the life time of this
universe is approximately $10^{17}$ years. And the size
of the universe is less than $10^{17}$ light year.   However,
the value of gravitational constant (G) is also decreased
$10^{34}$ times.

The image of the galaxies which are far away have
to be older than the closer ones.   The older galaxies
ought to be brighter and its M/L ratio (mass to
Luminosity ratio) has to be smaller.   However, the
result of actually observation is oppsite from the
above logic.   It is unreasonable to assume that older
galaxies contains more black holes and neutrinos.
Also, all attempts to detect a halo by its visual, inf-
rared, radio or x-ray radiation have failed.   However,
the older galaxies will have higher apparent mass
because the GRAVITATIONAL CONSTANT IS HIGHER
then.

IV. Symmetry creating mechanism.

What is the symmetry breaking mechanism? Or, what is the symmetry creating mechanism? Really, which one came first, symmetry or asymmetry? According to electroweak theory, the symmetry was broken after the big bang -- when the universe became cooler, the asymmetry appears. I would like to interpret this in a different way. I believe that symmetry was created during the big bang. Actually, before the big bang, this universe has never been degenerated. This universe was not symmetry because that the absolute time has one and only one direction ($\Delta t > 0$). But, during the big bang, the universe was degenerated. This absolute time degenerates to symmetrical time subcoordinates and 3 orthogonal space dimensions. At the same time, the super-heavy neutral particles began to decay. This universe was asymmetrical, and the symmetry was created.

Let's re-examine the defination of parity. Parity is an inversion transformation, which consists in simultaneously changing the sign of all the coordinates. For the cartesian coordinate, parity can be easily defined as two values operator ($p = \pm 1$). However, according to SUT, parity has to be a four values operator, $p = (+1, -1, +i, -i)$. From another view point, we always discuss the parity with the mirror image. But, we always use a flat mirror. It simply doesn't match with the reality. Our universe is bent because that the time is degenerated. We should use a curved mirror for defining the parity. And we will discover an amazing result. The image of an one ended arrow ( $\nearrow$ ) will look like a two ended and symmetrical arrow ( $\nearrow$ ) from a curved mirror. The absolute time has one and only one direction. It is asymmetry. But the degenerated time coordinates has bent this universe. Therefore, the image of this absolute time is symmetrical.

Also, the universe was created from "Wu" and it has to stay as "Wu" in God's view point (inf U). The "reality" (matter or anti-matter) is an image which was created by a supersymmetry operation. However, the present universe is asymmetrical. There is no anti- matter "naturally" existed in this present universe. But, this universe does possess a lot of symmetry properties (global and local gauge symmetry). In chapter IV, I have discussed how to represent our "nature space" by using a rigid gauge and symmetrical space. Here, I am going to discuss some examples how symmetry can be created from an asymmetrical vector. Let vector A = ( 1, 2, 3, 4, 5, 6, 7, 8, 9). Obviously, there is no special symmetrical properties can be found from vector A. Now, I am going to transfer vector A to a 3x3 matrix with following procedure.

i) Bending A $\longrightarrow$ B,

$$B = \begin{bmatrix} 1 & 2 & 3 \\ 4 & 5 & 6 \\ 7 & 8 & 9 \end{bmatrix}$$

ii) a. Rotating  B  $45^\circ$.

   b.  Rotating plane (159) & (753)  $180^\circ$.

iii) Stretching (move  4, 2 level with 9), then B' is,

$$B' = \begin{bmatrix} 4 & 9 & 2 \\ 3 & 5 & 7 \\ 8 & 1 & 6 \end{bmatrix}$$

Now, B' has a very good symmetrical properties. The sum for any column, row or diagonal is 15. Actually, this kind of symmetry can always be constructed by "bending", "rotating" and "stretching" an asymmetrical vector.

# CHAPTER SEVEN

## Colored numbers

The goal of this paper is to try to find a unique way to identify and represent all numbers. And I find that all numbers are colored. Furthermore, these colors have significant meaning in elementary particle physics.

I. Introduction.

What is a "number" and how can we define it? This is a well-answered question. However, re-examining its meaning, we may still find some fascinating results. We can define it as, "Every point on real line corresponds with a number". It doesn't matter how we define "numbers", we do need to find a way to represent them. A well-known system is by using ten digits ( 0, ..., 9). But, can we really represent all numbers by using these ten and only ten digits? If the answer is NO, than what else do we have to add into the ten digit number system?

II. Colors and digits.

For the reason to avoid confusions, I am introducing two concepts -- colors and digits. For a base 10 system, there are ten colors ( 0, ... , 9 ). For a number 125, it has three digits. The first digit on the right is a color 5. The second digit from right is color 2. Even if in the base 10 system, the representation of 125 is not unique. There are at least three ways.

1) 125 --- 3 digits.
2) 125.000.... --- infinite number of digits.
3) 124.999.... --- infinite number of digits.

From these three different representations I am very sure that the relationship between finite and infinite is very close. At here, I defined the digits of a number as the "SMALLEST" among its represen - tations. For example, 125 has three digits, not infinite.

In base 10 system, I have introduced 10 colors. A lot of numbers are the combinations of a few colors. For example, 125 is made of 3 kinds of colors. If we can designate a color for every number, then we will find out that we can't represent all numbers by using these 10 colors only. Every number can possess one color only. The defination is as follows, "The color of a number is the same color of the furthest digit to the right of a representation (in a decimal form) which has the least digits". For example, 125 is 5-colored with 3 digits. $1/3 = .333...$ is 3-colored with countable number of digits. $7/22 = .3181818...$ is 8-colored with countable number of digits. In this case, 18 is a repeated cell. The color of the cell is 8-colored. I defined the color of 7/22 by using its cell's color.

But, can we color code all numbers? Are 10 colors sufficient to represent all numbers? Very obviously, we are able to color code all rational numbers with 10 colors. However, we are not able to color code irrational numbers. Some new colors have to be introduced. For number such as $\sqrt{2}$, $\pi$, ... there is at least one additional color needed. I call it a plus color "+". For example, $\pi = 3.14159+$. Even with this additional color, the representation for $\pi$ is not perfect. Very clear, $3.14159+$ do not have to be equal to $\pi$ even if it does make sense to write $\pi = 3.14159+$. In base 10 system, every color is actually a number by itself.  IS COLOR "+" A NUMBER?

III. Countable, uncountable and binary system.
All rational numbers have at most countable digits.
1/3 = .333..., there are countable digits and it is 3-colored.  How many digits are in the number $\sqrt{2}$.  It
is the same as asking how many digits "+" color
contains.  We can make two guesses.
1) "+" color contains countable digits.
2) "+" color contains uncountable digits.
We can answer this by introducing a binary system.
Besides the base 10 system, we are able to represent
all numbers by using a base 2 system.  The base 2
system contains 2 colors only.  They are red = 0 and
white = 1.  Very clearly, all integer numbers can be
represented by these two colors.  However, we will
find some difference for fractional numbers.  In base
10 system, 1/5 = .2 is 2-colored with finite digits.
But, in base 2 system, 1/5 = .001100110011... We are
still able to identify 1/5's color by using the color of
its cell.  However, there are countable infinite of
digits for 1/5 in base 2 system.  There are only two
colors needed.  But, for digits, there is no way to
reduce the amount of digits of 1/5 to finite in base 2
system.  However, if we increase the number of
colors from 2 to 10, then 1/5 has only finite digits.
It is very reasonable to assume that countable infinity
is a direct consequence of eight missing colors.  So
far, we have not proved that every color is a number
even if some colors are numbers.  I would like to
give this new color a name --- color C (countable).
In base 10 system, there are 10 well-known colors and
one "+"-color.  However, color C is also a hidden
color in base 10 system. Are color C and color "+" the
same color?  Very obviously, there is a difference.
Now, we can redefine a color system in base 10 system

as follows,

1) For any finite digits numbers, its color is determined by the fartherest digits to the right. For example, .123 is 3-colored.

2) For any countable digits number, it has a C-color. For examples, $1/3 = .333... = .3C$ and $7/22 = .31818... = .3\overline{18}C$   $.3C$ & $.3\overline{18}C$ are very clearly defined.

3) For a irrational number, all of them are "+"-colored. Oviously, "+" color and C color are different. Now, let's make a summary for the above discussion.

i) For base 10 system, eleven or more colors are necessary unless we deny that irrational number is a number.

ii) For base 2 system, three or more colors are necessary just for rational numbers. For examples,

   $4 = 100$   red colored.

   $5 = 101$   white colored.

   $1/5 = .0011C$   C-colored.

But, what is "+" color? $\sqrt{2}$ and $\pi$ are both "+"-colored in base 10 system. However, there are some difference between them. The "+" color for $\sqrt{2}$ can be easily transfered to a C-color by squaring it. However, square $\pi$ does not effect the "+" color $\pi$ possesses. Therefore, I have to introduce two colors (color P and color "+") to represent them. Color P is a pseudo "+" color which is able to be transfered to color C by an algebraic operation.

For $\pi = 11.00100100011+$, the color "+" can not be removed by any means. Therefore, in base 2 system, at least five colors are necessary to represent all numbers unless we deny that irrational number is a number.

IV. Infinities, countable traps & pseudo-uncountable.

In the modern mathematics, there are two kinds of infinity --- countable and uncountable. However, I

have hinted that there are three kinds of infinity in section III. It is very easy to understand why all integers are countable. We can arrange them to a special order and construct a one to one mapping with natural numbers. Very obviously, set A is countable.
$A = N \cup \{0\} = \{1, 2, \ldots ,\infty\} \cup \{0\}$
We can prove it as follows.

$$1 \dashrightarrow 0$$
$$2 \dashrightarrow 1$$
$$\cdot \qquad \cdot$$
$$\cdot \qquad \cdot$$
$$\cdot \qquad \cdot$$
$$\infty \dashrightarrow \infty$$

However, if set B = A, but it has a definite order,and we can not rearrange it, then the only mapping is as,

$$1 \dashrightarrow 1$$
$$2 \dashrightarrow 2$$
$$\cdot \qquad \cdot$$
$$\cdot \qquad \cdot$$
$$\cdot \qquad \cdot$$
$$\infty \dashrightarrow \infty$$
$$? \dashrightarrow 0$$

Is set B countable? Actually, set B is constructed from two parts. The one is a countable trap "N", the other is a "0" which hides behind the trap. Practically, the counting procedure seems to have no way of getting out of the trap and to reach the element "0". However, this common sense may not be strong enough to prove that set B is uncountable.

It is reasonable to assume that the number of elements in N is $\infty$ (countable) and the number of elements in B is $\infty + 1$. If we can prove that $(\infty + 1)/\infty = 1$, then set B should be a countable set.

$$(\infty + 1)/\infty = \lim_{n \to \infty} (n + 1)/n = \lim_{n \to \infty} ((n + 1)/n)^{n/n}$$
$$= \left[(\infty + 1)/\infty^{\infty}\right]^{1/\infty} = e^{1/\infty} = 1$$

Very clearly, $(\infty + 1)/\infty$ is equal to 1.   But what are

$\infty^{\infty}$ and $(\infty + 1)^{\infty}$ ?   The  ratio  between them is

obviously not equal to 1.   Let $\infty^{\infty}$ = inf C,   and

$(\infty + 1)^{\infty}$ = inf P.    inf P should be larger than inf C.
But,   what is inf C?    If we assume that inf C is count-

able, then $\infty^{\infty} = \infty$ .   Again, what is inf P?   Does
inf P have same size **as** (0, 1)?  This question has been
stated as Cantor's continuum hypothesis.   He assumed
that the power of second ordinal number class is equivalent to the power of the continuum of real number.
Nowadays,  we all know that Cantor's continuum hypothesis does not possess a true-false value.        In the
previous  section,  we have experienced the existence
of a hidden color P.   It is reasonable to consider inf P
as a new kind of infinity.   For color P, it will become
color  C  if we change the base from 2 to 10.   Also, for
inf P,   it will become inf C if we lift the restriction of
set B.     It  seems to be existing in one kind of infinity
which  has a power between countable and uncountable.
I call this new infinity pseudo-uncountable (inf P).

V.  Colored numbers and physics.
    In section  III,  I have concluded that five colors are
needed to represent all real numbers in base 2 system.
However,  three  colors (C, P, +) and three infinities
are  closely  related.  (see chapter eight).   It is reasonable to make the following correspondings.
    Color C --- inf C (countable)
    Color P --- inf P (pseudo-uncountable)
    Color "+" --- inf U (uncountable)
    However,  they still have a fundamental difference.
Three  colors  (C,  P,  +)  were  derived from trying to

find a method to represent finite numbers, such as, 1/5, 1/3, ..., $\sqrt{2}$, e, $\pi$ , etc. But, the infinities were derived from sets. Colors are properties of "numbers". Are infinities also properties of numbers? If they are, then these infinities have to be colors too. If we assume that they are, then set Z has 8 elements (colors).

$$Z = \left\{ 1, \ 0, \ C, \ P, \ +, \ inf \ C, \ inf \ P, \ inf \ U \right\}$$

Now, there are two questions have to be answered.

1) Are infinities numbers (or colors)?

2) Are elements in set Z the most fundamental colors or are we able to reduce their number?

One color can be removed from set Z because of the following equation. inf C = 1/0. I choice to remove "0". (see chapter VIII). A new set Z' has 7 elements.

$$Z' = \left\{ 1, \ C, \ P, \ +, \ inf \ C, \ inf \ P, \ inf \ U \right\}$$

It may be difficult to realize the methematical significance about set Z'. However, it has very important physical meaning. In particle physics, 7 colors are needed to represent all elementary particles. These 7 colors are 4 quark colors and 3 generations. Furthermore, if our universe is going to expand forever, then there is only one phase existing at the END. When the universe reaches this point, the existance of two kinds of infinity (countable & uncountable) becomes unjustified. However, if this universe is going to bounce (cycle), then two kinds of infinity may be not enough. (see chapter V). However, for any thing that is finite, there are only four colors. How namy colors are needed to describe all genes (or DNA). Why do elementary particles have three generations but only the first generation is naturally existing? According to set Z', the particles of the 1st generation should have a finite life time becouse they are constructed from four and only four colors. When the universe goes into an infinitive phase (example, big bang), then

the second and third generation particles will appear.
The colored real line does have significant physics
meaning.   The   real   line   and   time-space   are   iso -
morphic.

VI. The set U.

$\pi$ and  e  are  both  "+"-colored  in either base 10 or
base 2 system.   Amazingly,  e  can  be  written as a
fractional  form  by using two infinities, e = inf P/inf C.
It  is  very  clear  that any "+"-colored number can not
have  a  denominator which has finite  number of digits.
Really, we are able to write all numbers in a fractional
form  r = p/q  if we  do not restrict p, q which have to
be  integers.   Example: $\sqrt{8}$ = 4/$\sqrt{2}$. $\sqrt{2}$ is     finite with
infinite  number  of digits.  However, is b = x/infinity a
number?  If  x  is finite,  then b is "0".  Very clear, b
is  a  number  even  if  x  is  not  unique  in the  above
equation.   The  infinities  may  not  act  the  same  as
finite  number,  however,  it  is  able  to induce many
numbers.   Such  as,  0,  e,  r  (gamma),  etc.  If we
really  want  to  deny  that infinities are numbers, then
at  least  they are number generators.  If I define set U
as follows, U = {all numbers exclude all finite numbers}
If  infinities  are  not  number,  then U is an empty set.
If infinities are number, then U has at least 3 elements
and it should be able to generate the following numbers.
 i)  x1 = c/infinity,   c is finite.
ii)  x2 = D/E,    D, E  are  both  infinities.  Then x2 can
be  either  finite or infinite.  Example, x2 =inf P/inf C
= e (finite).
iii) x3 = D - E,    D, E are  both  infinities.  x3 can be
either  finite  or  infinite.   Example,  x3 = D - E = r
(gamma) is finite.
We  all  know  that there is uncountable amount of "+"-
colored  numbers.   Therefore,  it  is  necessary to have
uncountable  amount  of infinitive numbers even  though

there are only three kinds of infinity.  (see chapter VIII).

So far, I have extend the real line.  However, I have not discussed any concepts such as compact, dense , open set, or perfect, but the only task is trying to give every number an unique representation.  I have not yet accomplished this goal even if I have classified all numbers into 7 colors.

## VII. Conclusion.

My simple goal, every number can be uniquely identified and represented, may never be able to be accomplished because that the existence of the infinities.  We are not only unable to clear identify and represent the number $\pi$ , but there is uncountable of infinitive number which can not even be observed.  However, the most important of all, the properties of real line is not only possesses mathematical meaning, but it is a fundamental law of nature.  All real numbers which are finite can be represented by four and only four colors.  However, if we include infinity in the real line, then seven colors are needed.  The elementary particles possess 7 colors, but all lifes can only have four colors in their gene!

# CHAPTER EIGHT

## Chromology

After introducing some color principles, it is very easy to prove that $\pi$ is a normal number. And $\pi$) has the highest order of infinity even if we can continue to construct a bigger diverging sequence forever.

### I. Introduction.

In chapter VII, I have introduced seven colors to represent all numbers. Four colors in the finite region are easier to understand. The three infinitive colors are difficult to imagine. Actually, they represent the degrees of the softness of nature. In the finite region, all numbers are relatively rigidly defined even if we can not exactly pin point $\pi$ with ten colors $(0, \ldots, 9)$. It is very clear that 2 is larger than 1 and $\pi$ is larger than 3.14 etc. However, for an infinitive number, plus 1 or minus 1 on it do not make any difference. The law of the excluded middle of logic and the law of contradiction also lost their controling power at infinitive region. We seem to be able to squeeze all infinitive numbers into a single point ($\infty$). In this chapter, I will begin with some color principles to probe the infinitive region.

### II. The set U.

I have introduced set U in chapter VII.
$$U = \{x \mid x = r), \quad r \text{ is a real number}\}$$
I called r) a B-number. It possesses a half bracket ). It is constructed with a finite number r but its decimal

62

point is removed. For example,

e) = 271828182845.............

1/3) = 33333333333333333333333333333333333333....

$\sqrt{2}$) = 141421356 .........

$\pi$) = 31415926538979323846263338327950288419 7....

Set U is uncountable since real numbers are uncountable. It is very clear that $\pi$ is larger than e. However, is $\pi$) larger than e)? Obviously, nobody knows.
Today, an universal accepted idea is that we can always
construct a sequence C which diverges faster than any
diverging sequence B. In chapter VII, I have hinted
that $\pi$) is large than $\sqrt{2}$) and $\sqrt{2}$) is larger than 1/3) and
$\pi$) has the highest order of infinity. In order to prove
this, some color principles are introduced in the next
section.


III. Color principles and definations.

In this section, I will use some undefined concepts
which are seats (digits) and a decimal point. However,
these concepts will become clear in the later sections
of this chapter. Here, I am going to use them to define what are finite and infinite.

Def. 1) A number r is finite if there are seats (digits)
on the right side of the decimal point.

Example, e = 2.71828182845... is finite because
there are infinitive number of seats on the right side
of the decimal point. r = 125 is also finite because
that r has a hidden decimal point and there are infinitive number of seats on the right side of it.

r = 125.0000000....

Def. 2) A number r is infinite if there in no seat (digit)
on the right side of the decimal point.

Example, e) = 271828182845..... does not contain a
decimal point, therefore, there is no seat beyond it.


Def. 3) S is a SELF if $S^S = S$. and there are four SELF.

$1^1 = 1 = S0$

$\inf C^{\inf C} = \inf C = S1$ (countable)

$\inf P^{\inf P} = \inf P = S2$ (pseudo-uncountable)

$\inf U^{\inf U} = \inf U = S3$ (uncountable)

Def. 4) p is a number if  i) $p = r) \times 10^{-S_i}$,  $r) \in U$,

or  ii) $p = r \times 10^{+S_i}$ , r is a real number.

Example,  $e = e) \times 10^{-S3}$,  $e) = e \times 10^{+S3}$

C-colored number has countable digits.  P-colored number has pseudo-uncountable digits.  "+"-colored number has uncountable digits.

Def. 5)  C is a color if $C(i) = 10^{-S(i)}$ , and $i = (0, 1, 2, 3)$
So, there are seven colors, $Z' = \{1, C, P, +, S1, S2, S3\}$ and four SELF.

Theorem 1) All numbers (finite or infinite) can be represented by seven colors or four SELF.

Proof : use Def. 4 and 5.

Theorem 2) The degrees of freedom for all numbers is four.

Principle 1) Color exclusion principle --- every seat can be occupied by one and only one color. Every number can possess one and only one color. On the other wards, colors are clearly seperated. "+"-color can never be able to become C-color or P-color and vice verse.

Theorem 3) Set $S = \{S0, S1, S2, S3\}$ , Set $C = \{1, C, P, +\}$ and topological sphere are topological isomorphic. (see chapter nine --- Unilogy).

Def. 6) Set S and set C are called black set or singular sphere or singular pole.

Def. 7) Color of a set is identified by the color of its element which is on the farthest right (the one contacts with the set symbol on the right side).
Example, $[0, 1]$ is white-colored (colorless).
$(0, 1)$ is C, P or "+"-colored.
Let, $[0, 1)c$ is C-colored.
$(0, 1)p$ is P-colored.
$(0, 1)+$ is "+"-colored.
Principle 2) Color complementary rule.

i) $[0, 1)c \cup [0, 1)p \cup [0, 1)+ = [0, 1]$

ii) $[0, 1)c \cup [0, 1)p = \overline{[0, 1)+}$ , etc.

iii) $[0, \infty)S1 \cup [0, \infty)S2 = \overline{[0, \infty)S3}$, etc.

Def. 8) Color neighborhood is a close set whose interior points are single colored, and its closure contains the complementary colors.
Theorem 4) Every number is surrounded by infinitive number of numbers which possess their complementary colors. In other wards, real line is quantized for colors.
Let, $C = \{x \mid x \in R(\text{real number}), x \text{ is C-colored}\}$
$P = \{x \mid x \in R, x \text{ is P-colored}\}$
$+ = \{x \mid x \in R, x \text{ is "+"-colored}\}$
Very clear, C, P, + are not connected sets.
Principle 3) Color force priority. Let $* = \{\oplus , \odot\}$
then, $1 * C = C$            $S0 * S1 = S1$
$C * P = P$        and      $S1 * S2 = S2$
$P * + = +$                 $S2 * S3 = S3$
The color force priority can be represented by another concept --- softness.
Def. 9) The softness index of a color is as follows,
$1 = 1 = S0$
$C = I = S1$
$P = II = S2$
$+ = III = S3$

Now, principle 3 can be rewritten as follows,

$$1 * I = I$$
$$I * II = II$$
$$II * III = III$$

IV.   $\pi$ is softer than 1/3.

$\pi$ is "+"-colored with softness III and 1/3 is C-colored with softness I.  Very clear, $\pi$ is softer than 1/3.  However, this also can be proved in geometry. In chapter IV, I have discussed the way to trisect an angle and quadrature of a circle.  An angle can be trisected by using a rigid space (compass and straight edge) with countable steps.  However, quadrature of a circle (singular pole) can not be accomplished by using any rigid space.  For an angle, it can be represented as follows,

$\alpha = 2 \pi /b$ (b is a real number which does not contain $\pi$ ).   Then,  $\beta = \alpha /3 = (.3C)( 2 \pi )/b$

Clearly, trisecting an angle is a C-colored operation and $\pi$ has not been decomposited.

For quadrature of a circle,  $\pi r^2 = a^2$.  Let r = 1, then   $a = \sqrt{\pi} = ( .56418958+) \pi$ .

Here, quadrature of a circle is a "+"-colored operation and $\pi$ has been decomposited.

There is not only a fundamental difference between trisecting an angle and quadrature of a circle in the mathematics view point, but it also has a significant meaning in physics.  Therefore, quarks have only 1/3 of charge.  The way of quadrature of a circle has quadurpled colors.  There are four quark colors (Red, Yellow, Blue and colorless) and four generations (generationless, 1, 2, 3).

V.  $\pi$ ) has highest order of infinity.

I can easily prove this by using colored mathematics.

However, I prefered to use a more orthodox method. There are two steps needed. The first is to prove $\pi$) is a normal number. I will discuss this in the next section. The second is to prove that $\pi$) has highest order of infinity. I will discuss the second first.

Let Y is the set with all diverging sequences.

$Y = \{x \mid x \text{ is a diverging sequence}\}$

For any $x \in Y$, $x(1)$, $x(2)$, ... , $x(n)$, ...

For every $x(n)$, we can always find a number m $(m \in N)$ and have the following relationship.

$$10^{m-1} \leq x(n) < 10^m , \quad n \in N$$

However, in the number $\pi$), we are always able to find a block of digit which contains $10^m$ of "0", and

this kind of block will appear every $(10)^{10^m}$ digits. There are infinitive numbers of this kind of block in $\pi$). This is possible because that $\pi$) is a normal number. Therefore, we can always find infinitive numbers of such block in $\pi$) to correspond with every $x(n)$. So, we can not find a sequence which has higher order of infinity than $\pi$) even if we can continue to construct a bigger diverging sequence forever. Actually all inf U-colored number has the same order of infinity.

VI. $\pi$ is a normal number.

Let $b = \pi - 3 = .141592654+$

$= .0010010000111111011010+$

Clearly, 0 and 1 are appearing with almost the same probability with this approximated calculation in base 2 system. If we can prove that there is no repeated cell in the binary representation, then b is a normal number.

Theorem 5) r(in binary system) has a repeated cell which contains q digits if r(in base 10) has n finitable digits on the right side of the deicmal point. And

$$(10)^{n-d} \leq q < (10)^n/2, \qquad d \leq n$$

Def. 10) r has n finitable digits after a decimal point if there are all zeros on the right side of the nth digits. If I can prove that q is uncountable, then b is a normal number.

STEP 1, $n \neq d$. If n = d then b has to be end with pure 0 or 1. In both case b is a rational number.

STEP 2, n is inf U. This is from the defination of b.

STEP 3, d can not be a inf U because of step 1.

STEP 4, if d is inf C or inf P, then

$$n - d = III - I = III$$

or $\quad n - d = III - II = III$

Therefore, $q \geq III$ (inf U) and b and $\pi$ ) are normal numbers.

VII. Cantor's continuum hypothesis.

Let N(0) represents aleph-null -- the first number class of cardinal number. And N(1) represents aleph-one --- the second number class of cardinal number. Cantor stated his continuum hypothesis with an algebraic form:

$$2^{N(0)} = N(1)$$

Very clear, N(0) is equal to inf C. If Cantor's hypothesis is correct, then N(1) has to be equal to inf U and inf P is not existing. However, with chromology, I can easily prove the existance of the inf P. In chapter IV, I have discussed the softness of nature by using quadrature of a circle. Now, I am going to discuss this from a different view point.

If a square's area is $4r^2$, then the biggest circle's area which is contained in square is $\pi r^2$. Let's paint the circle's area in red and the remainder of the square in white. The ratio ($K = red/4r^2$) is 78.54%. If we divide square into four equal smaller squares, and the total red area is unchanged by subdividing the original disc. And the ratio K is as follow when the above process continues.

$$K(n) = 2^{2n}( r/2^n )^2 \pi /4r^2 = 2^{2n} \pi ( r(n))^2/4r^2 = K$$

The above process produced some following results.
Theorem 6) When n = inf C, the number of discs are equal to,

$$2^{2\inf C} = 2^{\inf C} = 2^{N(0)}$$

Theorem 7) With $2^{N(0)}$ discs which has radius $(r/2^{N(0)})$ covers 78.54% of area of the square .

Very clearly, red color can never cover the square regardless how many discs have been generated. When n is finite, the red discs are very rigid and the above process is a D2 dense packing. When n goes into infinitive region, n = inf C and $r(n) = x(n) = (\sqrt{2}-1)(r/2^{2n})$, the red discs become soft and it is not a dense packing any more. We are able to insert $( 2^n + 1)^2$ pieces of white discs which has the same radius ( $r(n) = x(n)$) as the red discs to fill up the square. However, the number of the white discs are larger than the number of the red discs even if the total white discs occupies less area than the red discs. This is possible because white discs are softer than red discs. Red discs have an area which is equal to 0(hard) = red area/$2^{2N(0)}$ and white discs have an area which is equal to 0(soft) = white area/ $( 2^{N(0)} + 1 )^2$.

Theorem 8)

$$N(1) = ( 2^{N(0)} + 1)^2 white + 2^{2N(0)} red > 2^{2N(0)} red$$

In the infinitive region, the size of the number become impossible to compare, but it is very easy to distinguish them by using the colors. All logic developed in the finite region is no longer valid in the infinitive and soft region, and the chromology is the only valid logic which is able to probe the infinities.

# CHAPTER NINE

## Unilogy

Unilogy is a unified topology. It classifies the points into two categories --- geometrical point and black point. And, a ghost partner of our universe is residing in the black point. Therefore, all conservation laws remain valid during an unimorphic transformation --- from "NOTHING" to "SOMETHING".

## I. Introduction.

In mathematics, there are many kinds of space. Such as, Euclidean, Riemannian, topological space, etc. On the view point of super unification, there is one and only one kind of space. It is the "nature space". Therefore, all those different kinds of space must be equivalent to each other or approximately equivalent to each other. I remenber a story about an elephant. One man was interested by its legs and bone structure, therefore, he invented a rigid space. The other man was interested by its nose and ears, then he invented topological space. Obviously, from different ground rules and definations, we obtain different results. Here, I am only interested in the elephant itself as a whole. Immeadiately, I run into two major problems.

i) How to define what is an elephant. The defination and characteristics of legs and bone are quite different from nose and ears. Sometime, it is even contradicted to each other.

ii) Even if it is not impossible to make a perfect defination for a unified space, it is not very practical for trying to make a perfect defination in a few pages of paper.

Therefore, in this paper, all terms used will be explained by common sense. However, some common sense discussed in this paper may be very simple to understand for physicist and mathematician, but it is very difficult (not common) for other people. In other wards, this paper will not begin from a series of assumptions and definations. I will let the logic go wild by itself and let the "nature" to give the final verdict. Furthermore, in modern mathematics, we always used the law of contridiction to prove or disprove a theory or theorem. For example, "If A = B, then A $\neq$ B". It is clearly contridicted to each other and it does not possess a true-false value from the orthodox view point. However, do we (human) really have "THE RIGHT" to make a judgement for its true-false value? As soon as we get into the infinitive region, the law of the excluded middle and the law of contradiction become useless.

II. Unified topology -- black point and unimorphism.

In combinatorial topology, there has been many surfaces studied. Such as, sphere, disc, torus, crosscap, etc. In this paper, I am going to discuss a special transformation -- unimorphic transformation (UT). UT is performed by using two kinds of tools only. One is scissors. The other is thread. The thread possesses two characteristics. The thread head is very rigid (as a needle). Its body is very soft and can be squeezed into an infinitismal small ball (black point). If A can be transfered to B by UT, then A and B is unimorphic. Now, I am able to demonstrate that all surfaces (sphere, torus, etc.) are unimorphic. The following are some examples.

1) Torus to sphere -- the procedures are as follows,

  a) Cut torus open to form a cylinder surface.

b) Sew up both ends of cylinder.

c) Pull the thread to infinitive tie, then the seam becomes a point which contains infinitive many points. I call it -- black point.

The black point is different from all other points. All point on the torus is geometrical point, but black point contains infinitive geometrical point. From the above procedure, the "topological torus" is exactly equivalent to a "topological sphere" plus two "black points". In chapter VI, I have discussed one special characteristic of space. I called it -- "Wu". And

$$Wu = A - B$$

A is Euler characteristic. B is the total number of black points of a surface. Wu is the fundamental corner stone for Unilogy.

2) Cylinder to Mobius band,

a) Cut cylinder (alone z-axis) almost open (except one point).

b) Rotate one side $180^{\circ}$.

c) Sew it back up without tighten (no black point).

Here, I introduce a new concept "rotate". Does our nature allow this operation --- rotating. Or, does our nature possesses an angular momentum by itself.

3) Half sphere to a black point.

A "nature sphere" possesses two black points. Let's cut this nature sphere into two half-spheres. Mathematically, I can find a function to map the boundary edge (circle) to the black point (a sink). Actually, I can squeeze the whole half-sphere into the black point. Every physicist knows there is "CURL" for every sinking process. Therefore, the black point does possess an angular momentum.

So far, I have not defined what are surface, space, sphere, or even what is geometrical point. However, I did explain what is black point and its topological and dynamical properties. Black point is constructed by sewing up infinitive geometrical point together then squeeze them into a point which has a radius is equal to zero. During this process, an angular momentum is inplanted into black point.

4) The real line.

In chapter VII, I have introduced 7 colors for constructing the real line. Actually, the real line which not includes "infinity" can be constructed by 4 and only 4 colors. Actually, it is unilogically equivalent to a topological sphere. From this relationship, I have proved that 4 and only 4 colors are necessary to color the map on a sphere. However, for a real line which includes "infinity", it is unilogically equivalent to a "TORUS" and 7 colors are needed.

The black point which comes from by squeezing a half sphere is undistinguishable from a black point which comes from by squeezing a half-line.

III. Unified universe.

Now, I can explain the evolution of the universe from a black point. The steps are as follows.

i) "NOTHING" --- black point (see chapter VI).

ii) Stretch black point becomes a half-line with a velocity c(light speed). And give this line a name "TIME". This "TIME" has three very important characteristics.

a) It possesses angular momentum h.

b) It possesses linear momentum (the concept of mass appeared, $p = mc$ ).

c) It can be stretched to forever.

As soon as the "TIME" was created from black point, a "TIME CONE" appeared because of the angular momentum. Its projected view is an Archemede's spiral. This spiral trisected space. Charges and 7 colors appeared. The expanding force is time-like force. The forces induced from "TIME CONE" and

Archemede's spiral are color forces (space-like force). However, this expanding force can not be created just by itself. A ghost partner has to reside in the black point to maintain all conservation laws. The longer we stretch the time, the stronger the ghost will be and it will finally overcome the expanding force. Then this universe begins to contract back to black point (big bang or big crunch) and expands in the opposite direction (anti-matter). This universe is oscillating around the black point just the same as a pendulum. Therefore, the matter and anti-matter appear in each big bang alternatively. The total energy, linear and angular momentum are conserved for universe(s). However, the amplitude of its swing will double during each cycle.

IV. Conclusion.

This universe is expanding, but what kind of force is pushing it to move outward. The expanding force is UNILOGICAL force and the gravity is its ghost partner.

# CHAPTER TEN

## The philosophy of super unification

Physics describes nature. But, what is nature? Birth and death are natural phenomena. Why is there death? Some species reproduce through virgin birth. Virgin birth can reproduce twice as fast as any bisexual method. Furthermore, through virgin birth, the risk of distinction for species is a lot less than bisexual reproduction. Why does nature prefer bisexual reproduction? All lives struggle to reproduce. Why are all lives forced to reproduce? Are these religious questions? Can physics give any explaination for all questions listed above? The answer has to be "YES" if nature is able to be described. This philosophy is very different from the orthodox view point. It contains three principles.

I. Dynamical principle.

The universe is moving between two poles. The one is big bang. The other is the big crunch. The elementary particles also possess two poles (see chapter two and chapter five). All physical motions can be described by dynamical equation. And all dynamical equations involve a fundamental variable --- TIME. Time comes from nowhere and disappears automatically. What is time? There is an operational definition for time in physics. Physicists measure time by using events. They only know the quantity of measurement of time. They don't know what is time. And they don't even care to know.

We all know what is TIME. When we are waiting, time goes terribly slow. When we are happy, the good time always flies away with amazing speed. Our feelings are always effected by TIME. Our feelings came with birth. Even though the feelings can change, a life can not get rid of feelings. How feelings shape our life is a psychological problem. Why feelings exist in nature is a physics question. How can a physics law govern the process of the birth of our universe but independent from a natural phenomena -- feelings?

Artists are the people who can describe this nature, feelings, best. For a musician, all feelings can be represented by using seven notes. For a painter, all feelings can be represented by using seven colors. Music intrinsically involves time and there is no audience who is able to hear the second note before the first one is played. In contrast, there is no intrinsic "scanning order" built into a visual art that the eye must follow. Therefore, music is fundamentally one dimensional, but visual art are generally two- or three-dimensional. However, if a picture is three thousand miles long, then the scanning eye has to obey an order which is caused by our scanning ability. On cosmic scale, the visual art (colors) and music (notes) are essentially without any difference. There is not only existing a methematics proof that high-dimensional space can always be represented by a one dimensional variable, but biochemists are also found out that all lifes can be described with four colors or four notes (A, G, T, C). A is adenine. G is guanine. T is thymine. C is cytosine. Furthermore, human and yeast mitochondria have essentially the same set of genes, but their nucleotide sequence is different. In other words, all lifes are singing their own life

song with four universal notes or they are painting
their self-portrait with four universal colors.

Why are there only four notes (colors) for all lives?
Musicians invented seven notes. Are musicians
smarter than nature? Why do they never invent eight
or ten notes? These kinds of questions were meaningless before. However, we do know the answers now!
In chapter VII & VIII, I have proved that there are only
four colors needed for finite region. If a living cell
possesses five or more colors, then it will never die.
The predetermined fate of death is the direct consequence that all lifes have four colors only.

All lifes are struggling to reproduce the next generation. Is this the way all lives protest to the un-
fortunate fate of the death? Or is this the way how
nature expresses the meaning of the infinity? In
chapter VII, I have pointed out three additional colors
for infinities. For elementary particles, there are
three generations. And every generation has four
quark colors (see chapter II). All lives are also obey-
ing the same law. The female's egg and the male's
sperm are obviously one kind of life form which are
different from their producers even if they both contain four colors (notes). They both can not escape the
inevitable fate of death unless an unification process
take place between them (except an unisexual reproduction process). After a fertilization process completed, the second generation of life form begins. Do
lifes possess third generation -- inf U (eternity)?

II. Unobservable principle.

Is there life after death? The region of after death
is unobservable. The creation before the big bang is
also unobservable because there is nothing there to
observe or to be observed (see chapter VI). Unfortunately, all physicists are too proud to discuss this

kind of natural phenomena. A theory must make specific numberical predictions which can then be compared to experimental results. And all experiments need to have control and variable parts. However, the nature does not give us an absolute yard-stick. If all physicists in the world died simultaneously, the nature law will not hurt one bit. Einstein's special theory is the direct consequence of the topological property of the nature, and there is nothing to do with observers (see chapter IV & VI). His theory is valid but his interpretation is not exactly correct.

In chapter VI & VIII, I have discussed the infinities and softness. The softest spot (black point -- singular pole) is alway unobservable. Newton's equation , Einstein's theory both possess singular point because our nature does. However, we are able to see through the unobservable as soon as we realize its existence.

Newton realized that we don't really know how to measure a varing velocity, we don't really know how to measure the gravitational force between two irregular shaped objects, we don't even know how to analyze a curve. The only things we know are the straight line, constant velocity, etc. Therefore, Newton cutted all curves into infinite number of pieces. Those small pieces will be straight enough for us to handle. He invented Calculus. He turned our inability into a wonderful tool to discuss this complex universe.

Heisenberg realized that we are never able to locate a quantum particle's position exactly and knowing where it is really going at the same time. The human ability seems to be so little and so helpless. But as soon as he discovered the uncertainty principle, we realized immediately that the electron is unable to stay in the nucleus. And then, we developed quantum mechanics. Uncertainty principle is trying to prevent

us  to see through the secret of the micro-world.  But,
as  soon as we accepted uncertainty principle, we con-
quered micro-world.

Einstein  realized that we did not inherit an absolute
yard-stick.  We can not measure any distance correct-
ly  when  we  (us and measured object) have a relative
velocity.   We  can  not  even  find  out  the truth of an
event  regardless how many observers we have.  They
always  report  differently.   However,  as soon as he
realized  our  handicap,  he discovered the special relativity  theory.   And  then  there  is not any relative
motion  which is able to confuse us.  Furthermore, he
realized that we can not even distinguish the difference
between  $m(A)$ and $m(G)$.  And then, he discovered the
general relativity theory.

What  kind  of handicap do we have but we still never
realize yet?  Our most serious handicap is that we are
never  able  to  see  the  whole secret of our universe.
Our  universe has its own absolute meaning, value and
behavior.   The unobservable is the only nature of our
nature.   This universe does not provide us an absolute
yard-stick.   Our think ability is belong to this universe
and is never able to escape beyond it.

III.  Supersymmetry principle.

In  chapter  VI,  I  have mentioned that the material
world  is  not  symmetrical  even  if  any  degrees  of
symmetry  can  be  constructed.  However, the supersymmetry  is  existing  at  ETERNITY  and  Wu  state
(see chapter V).   The creation of universe is the same
as  the  creation  of  jobs  through  federal government.
When  Uncle  Sam  creates  jobs  from "NOTHING", a
ghost  partner  (DEFICIT)  is  also  created.  When more
bonds  was  sold  to  refinance  the  deficit,  the  ghost
grows  big  and  bigger.   However,  the  higher  the
deficit,  the  more  likely to increase the tax.  Soon or

later, the budget surplus (anti-deficit) will appear.
After long periods of surplus, the GNP will begin to
contract, and a new job program is needed. Therefore, deficit and surplus are alternatively appeared
during each economical cycle. The matter and anti-
matter are also alternatively appeared during each
big bang. And, this is the only way that it is possible
to create "SOMETHING" from "NOTHING". (see
chapter VI).

However, this supersymmetry principle is not only
demonstrated with economics and the universe (there
is no anti-matter naturally existing in this universe),
but it is also clearly demonstrated by lifes. Most of
the lives reproduce through a bisexual reproduction
which is merely logical and economical. The unisexual reproduction (partenogenesis or virgin birth) could
reproduce more than twice as numerous as the bisexual method for the reason that all the unisexual species
deposit eggs whereas only 50 percent of the bisexual
species do so. Furthermore, the unisexual reproduction method has zero risk for the possibility to waste
eggs for the reason that the unification with a sperm
is not guaranteed. Why nature has given up all these
advantages of virgin birth but choose a highly risky
method (bisexual) to reproduce. For a bisexual reproduction process, an ovarian cell that is DIPLOID
        has two sets of chromosomes. However, this
ovarian cell will become a functional egg which is
HAPLOID and contains only one set of chromosomes.
For this, the egg lost half of the genetical message
and it is not the same kind of life form as its producer.
It is very clear that the first generation (haploid state)
and second generation (diploid state) of life form are
completely different. There must have some fundamental forces which are pushing the nature to follow
this risky and complex method for reproduction.

Actually, there are two reasons. The one generation (unisexual species) life form and two generations (bisexual species) life form are allowed to exist because that there are three generations in nature. Furthermore, the bisexual reproduction is prefered because that the nature is supersymmetrical as a whole even if it is asymmetrical locally. In chapter nine (Unilogy), the universe was created from a "black point" (NOTHING) and it will return into a black point. The black point is the result of unification of matter and its ghost partner (anti-matter). The life is also coming from a "BLACK POINT". The unification of a haploid egg and its partner ( a haploid sperm) has created life. The bisexual reproduction is the direct proof for the supersymmetry principle and the virgin birth is the proof for the existence of the local asymmetrical property of nature.

IV. Conclusion.

If we don't believe in God, then where does our physics laws come from. If God does exist, then how can we reject Him from all of the physics theory and still hope that theory is telling the full truth? The primary question is that, can we imagine some ideas which are not permitted by God's law. We can prove something. We can disprove something. Can we always be able to do both or at least one of them? The question of God is outside the realm of physical experiments. Any experiment needs to have control and variable parts. God can not excluded from either of these. Physics discusses those aspects of the world which seem free from detailed divine intervention. Does physics really only discuss those aspects of the world which free from detailed divine intervention? Can every secret of this universe be verified in our laboratory? Is human in the center of our nature law,

or  at least plays part of the rules?  What is the relationship  between the nature law and our experimental
laboratory? What is the relationship between the nature
law  and  our  human life and spirit?  Is this universe
containing  only  pure  science  without  any  kind  of
mystery?  Is  the  interplay  between  experiment and
theory  the  only  correct way for understanding of the
laws of nature?  Actually, to accept the concept of God
is the key point in this book.

What  is  religion?   What  does religion talk about?
The religion has to discuss the followings.

1) The  creator -- we call Him God. (an unobservable
singular point).

2) After death (another unobservable singular point).

3) Earth life. (dress code, worship code, church code).

The  difference between science and religion is that
science  does  not  willingly  discuss  the  creator and
after  death.   Also,  all  sciencists are intentionly to
avoid the issues.  If you are interested  in theormody-
namics  only,  you  do  not  need  to  worry  about  the
creator.  If you are interested  in classical mechanics
only, you don't need to worry about after death.  If you
are  interested  in electronics only, you don't need to
worry  about where the haman mind comes from.   But,
if  we  are  interested  in finding a nature law, we are
better not ignore or avoid any of the aboves.  Einstein
said,  "the  nature  law  has  the same formular in all
frames".  I can give it a better defination.  The nature
law  is  everywhere.   It exists in your lab regardless
whether  you  can  find  it or not.  It exists in a pile of
hay regardless how much we have ignored it.  It exists
in  our  heart  regardless  whether we admit it or not.
God  is  highest  "SELF".  He is unobservable.  He is
supersymmetrical  and  dynamically  moving  between
creation  and  eternity.   His image is everywhere -- in
mathematics  (Chromology),  in physics (QCD) and in
lifes.

# A PERSONAL NOTE

Twenty years ago, I was in the 10th grade. I read an introductory book about the special and general relativity theory. I really enjoyed Einstein's wild imagination and super orgnized logical thinking. But, his faith of finding a unified field theory was the most powerful force ever to strike my heart. I knew that this universe is unified. I also knew that the time is a quanta. But, I just simply didn't know how to explain it. I couldn't put any of my idea into any kind of formula. Every time I tried to image how this universe really works, I can only see a huge black cloud floating in front of my eyes. However, I did see some sparks break out of these dark clouds once in a while. Unfortunately, they always disappeared even before I wanted to catch them. I almost gave up the search for a unified theory even if my faith for it remained strong. Fifteen years of disappointment and agony was a long time for a twenty-nine years old young physicist. However, after fifteen years practice, I had a habit of allowing my mind to float into my dark universe clouds when I take a rest. On December 4 , 1979, I finally caught the sparks which has teased me for fifteen years. I wrote my first paper which I called --- An assumption. My idea was so crazy, and Dr. Robert Mills wrote to me, "Do think carefully about what you can do best (not physics) and devote your energy to that". This is really the best advise I ever had. And I wish to thank all the people who gave me advice. The following is a name list of all my friends. Furthermore, some of their letters are attached as a historical documentation. And it demonstrates much of the struggle I have been through before this book was possible for publication.

# ACKNOWLEDGMENTS

Special appreciation goes to all of those who are listed below.

A. Dalgarno, Smithsonian Astrophysical Observatory.
John Schmidt, Princeton University.
Gart Westerhout, United States Naval Observatory.
Gordon Dunn, University of Colorado.
D. Wayne Cooke, Nemphis State University.
L. E. Poorman, Indiana State University.
Lawrence Puckett, U. S. Army Armament Research.
Lyman Mower, University of New Hampshire.
R. R. Newton, The Johns Hopking University.
David Finkelstein, International Journal of Theoretical
                    Physics.
Brian Matthew, University of Oregon.
David Axen, University of Victoria.
C. W. Tittle, Southern Methodist University.
Philip C. Keenan, The Ohio State University.
Sally Smith, Gravity Research Foundation.
R. Lee Ponting, Bennett College.
Mary Shoaf, Princeton University.
Jay Burns, Florida Institute of Technology.
Charles Nissim-Sabat, Northeastern Illinois University.
R. G. Palmer, Duke University.
C. K, Reed, National Research Council.
Daniel Galehouse, University of Akron.
Sumner Davis, University of California, Berkeley.
John Cooper, Naval Postgraduate School.

BERKELEY · DAVIS · IRVINE · LOS ANGELES · RIVERSIDE · SAN DIEGO · SAN FRANCISCO  SANTA BARBARA · SANTA CRUZ

DEPARTMENT OF PHYSICS                    BERKELEY, CALIFORNIA  94720

January 22, 1980

Jeh-Tween Gong
553 E. Front Street
Logan, Ohio  43138

Dear Jeh-Tween Gong,

Thank you for your letter of January 17, 1980, reminding me about not responding to your earlier correspondence. I am sorry but we do not have anyone here who has time to read your material, which we are returning to you herewith.

Yours sincerely,

J. D. Jackson
Chairman

JDJ:fs

Enclosure

Dr. Jeh Tween Gong
P.O. Box 73
Marshall, IL 62441

Dear Dr. Gong:

As requested in your letter of 30 June 1980, your manuscript,
"An Assumption," has been reviewed by the Naval Research Laboratory.
As you know, the task you took upon yourself, namely "to drive all the
physics forces (gravitational, electrical, and nuclear) from one
assumption" was too difficult even for Einstein.

The main objection to your letter is the assumption made in page 1,
Section I, iv, in which you give a particle with rest mass $m_0$ a velocity
$C = 3.10^8$ m/sec. According to the special theory of relativity you need
infinite amount of energy to do so. This assumption is not valid.

For the last hundred years, physicists have been building a con-
vincingly beautiful model to represent the world around us. One of
the main cornerstones of this building is the special theory of relativity.
The other one is the conservation of energy and momentum. Your assumption
seems to disregard those cornerstones. There are so many experiments
that prove the correctness of the special theory of relativity that it
would be hard to convince any physicist that he should abandon the
special theory of relativity for the sake of a unified theory.

Sincerely,

ALAN BERMAN
By direction

Southern Illinois
University at Carbondale
Carbondale, Illinois 62901

Department of Physics and Astronomy
(618)-453-2643

August 22, 1980


Dr. Jeh-Tween Gong
P.O. Box 73
Marshall, Illinois 62441

Dear Dr. Gong:

Thank you for sending me the manuscript. It would certainly be good to have a unified theory.. as you know, Dr. Salam, Weinberg and Glasow made the first step towards it and unified weak and electromagnetic interaction and were awarded the Nobel prize last year. Regarding your theory, I fail, however, to understand your supplemental assumptions.

For a particle at rest "c" $\neq$ v because that will violate many thing in special relativity. In fact if v = c, the particle cannot have rest mass.

Because your theory is based on these assumptions, which are not supported by experimental evidence. I suggest your re-examining it.

Sincerely yours,

F. Bary Malik
Chairman


FBM/nn


p87

# CALIFORNIA INSTITUTE OF TECHNOLOGY

CHARLES C. LAURITSEN LABORATORY OF HIGH ENERGY PHYSICS
PASADENA, CALIFORNIA 91125

October 6, 1980


Jeh-Tween Gong
P.O. Box 73
Marshall, IL 62441

Dear Jeh-Tween:

   I am sorry not to be able to spend any time considering your ideas, for I am working on quite different matters.

Sincerely,

Richard P. Feynman


RPF;ht

p88

P. O. BOX 5300
SYDNEY, CAPE BRETON
NOVA SCOTIA
CANADA
B1P 6L2
(902) 539-5300

March 26, 1982

Mr. Jeh-Tween Gong
P.O. Box 26294
Trotwood, Ohio     45426

Dear Mr. Gong:

   I regret by belated answer to your letter of inquiry relevant to my professional assessment of your paper. I regret further that illness has prevented my usual secretary from advancing the mail flow from the depths of a true avalanche of paper.

   Again, I was out of the country during parts of both the months of February and March, 1982. Your paper and the inquiries made have perchance been directed to the wrong individual. Whereas my background was in physics, my specialty was in radio physics and microwave propagation. It is now some 12 years since I became involved in College Administration as Dean of Arts & Sciences. Frankly, in respect to my proficiency to judge adequately your presentation, my physics is extremely rusty, my grasp of unified field theories very marginal. I feel, therefore, it would be very ill-advised for me to pose as an expert in respect to your paper. Perhaps you should redirect your inquiry to some of the excellent physics departments in the U.S. which have ongoing active research programs in theoretical physics.

                    Yours sincerely,

                    Dr. W.M. Reid
                    Academic Vice-President

WMR/agg

**OAK RIDGE NATIONAL LABORATORY**
OPERATED BY
UNION CARBIDE CORPORATION
NUCLEAR DIVISION

POST OFFICE BOX X
OAK RIDGE, TENNESSEE 37830

April 1, 1982

Dr. Jeh-Tween Gong
P. O. Box 26295
Trotwood, Ohio  45426

Dear Dr. Gong:

Thank you very much for sending us a copy of your "Unified Theory". Some of the concepts you presented were very thought provoking to my staff and me. Unfortunately, we are unable to provide a very good review of your work because we are mostly engineers and are not very conversant in the subject of unified field theories. I would suggest that a university physics department might be a more appropriate place for you to seek good professional reviews of your work.

My technical assistant has sent a copy of your paper to another part of the Laboratory in which several capable physicists are working. Since these people do not work under my supervision, I cannot say if and when they will respond.

I would like to wish you the best of luck in your work.

Sincerely,

*D. B. Trauger*

D. B. Trauger
Associate Director for Nuclear
and Engineering Technologies

DBT/JDW/cah

DEPARTMENT OF PHYSICS
**SHOCK DYNAMICS LABORATORY**
(509) 335-7217

November 8, 1982

Dr. Jeh Tween Gong
P.O. Box 26294
Trotwood, Ohio    45426

Dear Dr. Gong:

Thank you for your paper on unified field theory.  I do not feel qualified
to comment on this subject, and have passed it on to Dr. James Park of this
department.  If he has comments, he will address them to you directly.

Very sincerely yours,

George E. Duvall
Professor of Physics

GED/dm

cc:  J. Park

Reply to Attn of  103                                    December 17, 1982

Mr. Jeh-Tween Gong
P. O. Box 26294
Trotwood, OH  45426

Subject:  Unified Field Theory

No one at NASA Langley is involved in unified field research,
and therefore we cannot comment on your ideas.

To assess its worth, I recommend you consider submission of
your work to a peer-reviewed journal.

*Ed Prior*

Edwin J. Prior
Technical Assistant
   to the Chief Scientist

**DEPARTMENT OF THE ARMY**
ARMY MATERIALS AND MECHANICS RESEARCH CENTER
WATERTOWN, MASSACHUSETTS 02172   GASSNER/dac/ 617-923-5521

DRXMR-PP

7 April 1983

Mr. Jeh-Tween Gong
P.O. Box 26294
Trotwood, OH  45426

Dear Mr. Gong:

This is in response to the copy of your "Unified Theory" that you
submitted to our Director, Dr. E. S. Wright.

The Army Materials and Mechanics Research Center (AMMRC) is charged
with the responsibility of conducting applied research, exploratory
development, and advanced development efforts in the area of structural
materials and the mechanics of materials.  As such, we are not involved
in evaluating, nor do we have the necessary expertise to evaluate,
innovative quantum mechanical theories which attempt to explain
fundamental problems dealing with time and space.  It is suggested that
you forward your manuscript to a "refereed" academic journal in an
appropriate area.  They should be able to provide more meaningful
comments on your manuscript.

Thank you for your interest.

JOHN J. GASSNER
Chief,
Program Planning Division

**Taylor University**

Upland,
Indiana
46989

April 29, 1983

Mr. Jeh-Tween Gong
P.O. Box 26294
Trotwood, Ohio 45426

Dear Mr. Gong:

Thank you for sending me a copy of your paper entitled, "Grand Unified Theory". I particularly enjoyed the personal note on page 12.

You ask for my opinion regarding the paper. Unfortunately I am not qualified to serve as a reviewer on that topic. However, I wish you continued success as you develop the theory.

Sincerely yours,

*Elmer Nussbaum*

Elmer Nussbaum
Professor of Physics

# STANFORD UNIVERSITY

STANFORD LINEAR ACCELERATOR CENTER

*Mail Address*
SLAC, P. O. Box 4349
Stanford, California 94305

July 7, 1983

Dr. Jeh-Tween Gong
P. O. Box 26294
Trotwood, Ohio 45426

Dear Dr. Gong:

We shall be happy to consider you for a research appointment in the Theory Group of the Stanford Linear Accelerator Center for the coming year. I would encourage you to complete your application by having your professors send letters of recommendation in support of your application. If you have any supporting documents you wish to submit, we shall consider these also.

Applications will be reviewed early in January, and we shall let you know the results as soon as possible.* Meanwhile, I shall be happy to answer any further questions you may have about what goes on here at SLAC.

Sincerely yours,

Sidney D. Drell

sj

* It has long been SLAC's policy to put no deadlines on acceptance of its offers of RA's. We also, on occasion, have delayed offering an RA until one of our current RA's has successfully location an appropriate position at the end of his or her appointment. Therefore, it is not uncommon for us to still be offering Research Associateships as late as April.

# Receipt and Publication Charge Certification

The Editorial Office of Progress of Theoretical Physics acknowledges receipt of the paper you submitted for publication. We make it a principle to publish papers with publication charge, so we would like to confirm you the payment of the publication charge before accepting your paper to our journal. In case your institution or you cannot pay it, your paper will not be published in Progress of Theoretical Physics.

Please fill the form below and return it as promptly as possible to the Editorial Office.

Editorial Office
Progress of Theoretical Physics
c/o Yukawa Hall
Kyoto University
Kyoto
Japan 606

Title of Paper :

### Grand Unified Theory

Authors : Geh-Tween GONG

Received Date : August 15, 1983

Ref. No. : 3820

(Please quote for further communication.)

---

# Receipt and Publication Charge Certification

The Editorial Office of Progress of Theoretical Physics acknowledges receipt of the paper you submitted for publication. We make it a principle to publish papers with publication charge, so we would like to confirm you the payment of the publication charge before accepting your paper to our journal. In case your institution or you cannot pay it, your paper will not be published in Progress of Theoretical Physics.

Please fill the form below and return it as promptly as possible to the Editorial Office.

Editorial Office
Progress of Theoretical Physics
c/o Yukawa Hall
Kyoto University
Kyoto
Japan 606

Title of Paper : Grand Unified Theory —— Color Force and
Particle Decay

Authors : Jeh-Tween GONG

Received Date : July 18, 1983

Ref. No. : 3727

(Please quote for further communication.)

p96

# THE PHYSICAL REVIEW

AND

# PHYSICAL REVIEW LETTERS

EDITORIAL OFFICES  1 RESEARCH ROAD
BOX 1000  RIDGE NEW YORK 11961
Telephone (516) 924-5533
Telex Number 971599
Cable Address  PHYSREV RIDGENY

25 October 1983

Mr. Jeh-Tween Gong
P. O. Box 26294
Trotwood, Ohio 45426

Dear Mr. Gong:

We have received your manuscripts entitled "Grand unified theory --Color force and particle decay", "Grand unification -- Topological representation", "Our universe is closed", and your most recent letter entitled "Comparison of grand unified theories -- Introduction of topological GUT".

We regret to inform you that the subject matter in these papers is outside the scope of the Physical Review. Such highly speculative material as presented in these papers, and in your earlier paper entitled "Grand unified theory" that we returned to you last June as unsuitable for the Physical Review is more appropriate for those journals specifically devoted to the foundations of physics and specializing in speculative theoretical ideas. We have made no judgment on whether your work is correct or not, only that the subject matter is not suitable for the Physical Review.

We are returning your manuscripts.

Yours sincerely,

D. Nordstrom
Physical Review D

DN:cp
enc.

888 SEVENTH AVENUE, NEW YORK, N.Y. 10106/212-262-7990

10/31/83

Dear Mr.Gong:

Thank you for sending us your
papers for review.

Naturally, the material is much
too advanced for our use, so
we are returning the papers to you.

We appreciate your interest
in SCIENCE DIGEST.

Sincerely,

Frances Bishop
Editorial Asst.

A PUBLICATION OF HEARST MAGAZINES, A DIVISION OF THE HEARST CORPORATION

p98

Editorial Offices

Department of Physics
University of Illinois
  at Urbana-Champaign
1110 West Green Street
Urbana, Illinois 61801

Telephone
(217) 333-6859
(217) 333-0115

Associate Editors
D. E. Baldwin
E. D. Commins
R. W. Keyes
E. H. Lieb
C. Quigg
T. M. Rice
E. E. Salpeter
R. H. Silsbee
G. K. Walters
J. Weneser

Assistant to the Editor
K. Friedman

Published by
The American Physical Society
through
The American Institute of Physics

## Reviews of Modern Physics

Editor

David Pines

November 30, 1983

Mr. Jeh-Tween Gong
P.O. Box 26294
Trotwood, Ohio  45426

Dear Mr. Gong:

Thank you for your submission to Reviews of Modern Physics
entitled "Color forces, particles decay, and QED." I regret to
inform you that we cannot consider your paper for publication
because our journal does not publish original research.

As our name implies, we specialize in review articles, which
place the topic under consideration in historical perspective,
summarize the recent progress, discuss future directions that research is likely to take, and provide a comprehensive bibli-
ography.  These reviews are intended to keep physicists informed
of the work being done in important areas of physics outside
their own specialties.

Clearly your manuscript does not fit the above description.  It
belongs in the original literature rather than in a review
journal.  I am therefore returning it to you with best wishes for
your finding a more appropriate publisher.

Yours sincerely,

David Pines

DP/sd
Enc.

## THE PHYSICS TEACHER

Clifford E. Swartz, Editor   o   Thomas D. Miner, Associate Editor
Lester G. Paldy, Consulting Editor   o   Arthlyn Ferguson, Managing Editor

*Published for the American Association of Physics Teachers by the American Institute of Physics*

DEPARTMENT OF PHYSICS   •   STATE UNIVERSITY OF NEW YORK   •   STONY BROOK, NY 11794   •   516-246-6058

July 8, 1982


Mr. Jeh-Tween Gong
P. O. Box 26294
Trotwood, Ohio 45426

Dear Mr. Gong:

   Our journal is not concerned with physics research.  Neither readers **nor**
reviewers would be capable of analyzing your article on a unified theory.
This sort of work should be submitted to research journals such as *The
Physical Review*.

Sincerely yours,

Clifford Swartz
Editor


CS/nd

p100